ECHO MAKER

ECHO MAKER

Craig Macdonald and the Lives That Produced One of Canada's Most Significant Historical Maps

JAMES RAFFAN

Publisher and acquiring editor: Meghan Macdonald | Editor: Susan Fitzgerald
Cover designer: Laura Boyle
Cover image: Crane: istock/John_Wijsman; Map: Craig Macdonald

Library and Archives Canada Cataloguing in Publication

Title: Echo maker : Craig Macdonald and the lives that produced one of Canada's most significant historical maps / James Raffan.
Names: Raffan, James, author
Description: Includes bibliographical references.
Identifiers: Canadiana (print) 20250224895 | Canadiana (ebook) 20250224909 | ISBN 9781459755765 (softcover) | ISBN 9781459755772 (PDF) | ISBN 9781459755789 (EPUB)
Subjects: LCSH: Macdonald, Craig K., 1946- | LCSH: Cartographers—Canada—Biography. | CSH: Temagami, Lake, Region (Ont.)—Maps. | CSH: Cartography—Ontario—Temagami, Lake, Region—History—20th century. | CSH: Traditional ecological knowledge—Ontario—Temagami, Lake, Region. | CSH: Indigenous peoples—Ontario—Temagami, Lake, Region. | LCGFT: Biographies.
Classification: LCC GA473.7.M33 R34 2025 | DDC 526.092—dc23

We acknowledge the support of the Canada Council for the Arts and the Ontario Arts Council for our publishing program. We also acknowledge the financial support of the Government of Ontario, through the Ontario Book Publishing Tax Credit and Ontario Creates, and the Government of Canada.

Printed and bound in Canada.

Dundurn Press
1382 Queen Street East
Toronto, Ontario, Canada M4L 1C9
dundurn.com, @dundurnpress

To Richard and Victoria Grant
and all keepers of the reconciliation fire.

Forgetting is a result of colonization.
Craig Macdonald was on our path.
We are thankful.

— Chief Shelly Moore-Frappier,
Temagami First Nation

Contents

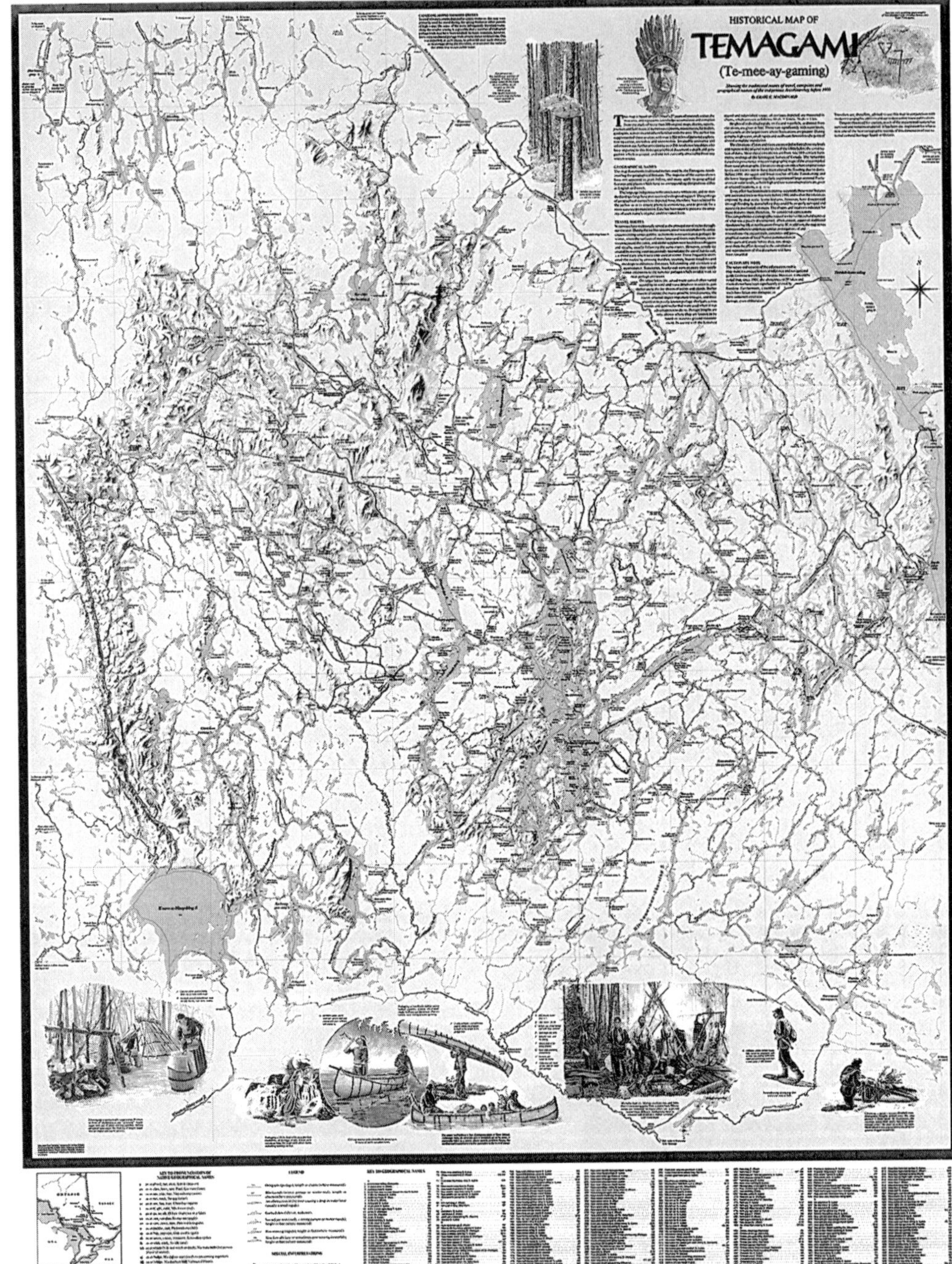
HISTORICAL MAP OF
TEMAGAMI
(Te-mee-ay-gaming)

Positionality Statements

From the Temagami First Nation (TFN)

In the spring of 2020, Victoria Grant came to the Chief and Council to inform us of the book project that she and Richard Grant proposed to sponsor through the Temagami Community Foundation. Their intent was to tell the story of Craig Macdonald and the development of the map. She was looking for an expression of support from the community. She informed us that they had a well-regarded author, James Raffan, who would work with Craig on the manuscript. She said that the author and Craig were going to come and spend some time in the community and speak with members. James would do some interviews to better understand the story and its context.

She also informed us that the book would be paid for through funds that she and Richard were donating to the Temagami Community Foundation. The manuscript would be owned by the foundation and the royalties from the book would go to the foundation into a fund that would benefit the Teme-Augama Anishnabai and n'Daki Menan. Victoria and Richard would play a lead role in the project. Council recognized that it is important that this story be told while Craig is still able to contribute to the telling. At that Council meeting in 2020, the Council agreed to support the project.

James and Craig spent some time in the community and had many conversations with members. They also spent a day with Michael Paul, one of the Council members, in a small settlement in Austin Bay. James made many visits to n'Daki Menan throughout 2021.

It was important for us to be part of this project. The book and the map are about our homeland. It is through the voices of our Elders that Craig was able to do this work. We honour Craig because we also understand that without his expertise and ability, the respect with which he approached our Elders and earned their trust, the time he spent to learn from them much of our language, and his personal commitment to seeing it through to completion, this map would not have happened. Like our Elders, he is also an extraordinary man who, with the support of his family, has given us something priceless. We don't know of any other First Nation that has access to this kind of information on their homeland. What we do know is that all lakes and rivers, the portages and winter roads everywhere across this country have a similar history. They are all homelands to the original people.

As we move forward with the project, we will ensure that the interests of the Teme-Augama Anishnabai are respected and protected.

From the Temagami Community Foundation (TCF)

In 2020, the Temagami Community Foundation was approached by Victoria and Richard Grant to sponsor the project, a book about Craig Macdonald and his remarkable map of the Temagami area.

Canada is engaged in a process of Truth and Reconciliation, and this is one story that demonstrates and illustrates the deep, rich history of the Teme-Augama Anishnabai, one of the many Indigenous Peoples in this country. We happen to live here on their traditional homelands.

The TCF has among its objectives "to promote the well-being of all members of the Temagami community by supporting cultural, artistic and educational activities." We believe all members of our community will benefit from this book. The board was excited to be part of this important project.

In response, the Directors passed the following motion:

> Resolved that: the Directors approve the participation by the TCF in the project outlined to the Board at a meeting May 25, 2020.

The TCF looks forward to the book being published and to see its distribution begin.

From Victoria and Richard Grant, Facilitators for the TFN and TCF

We have known Craig, and his map project, since the 1970s. In 2008, at an event at Bear Island, Craig made a presentation to the community in which he told some of the story of how the map came to be. It was a deeply moving experience for all of us who were there. Craig was returning something priceless to the Teme-Augama Anishnabai — knowledge he had acquired from our Elders and was bringing back with admiration and respect. And the community assembled that day understood and were generous in expressing their gratitude.

We asked Craig if he would consider telling the story of the map in a book. He said he didn't have it in him, but if someone wanted to write it, he certainly would support the effort. We were both committed to making this happen. And from then until 2020, we considered and approached authors who we felt might be interested in working with Craig and be comfortable in the community, but we never found anyone until 2020.

That year, owing to a remarkable series of circumstances, we found that author.

Craig, *Meegwetch*. There are not words in English or Anishnaabemowin that can express what we feel you have done by creating this map and preserving these stories.

James, *Meegwetch* to you as well. We cannot think of anyone else who could have captured Craig's story with such elegance and respect.

We acknowledge and thank the chiefs and councillors, and community members of the Temagami First Nation and the Teme-Augama Anishnabai. All of us involved in this project, and particularly Craig, James, and the two of us, are acutely aware that we are telling a story based upon information and content that belongs with you.

We acknowledge and thank the directors of the Temagami Community Foundation who agreed that this story was important to the constituencies it serves, the permanent and seasonal residents of the Temagami area, and the Teme-Augama Anishnabai. The Temagami Community Foundation is the sponsor of this project and has supported it from the outset.

We contributed the funding that created this manuscript because we think what Craig has done is astounding and important. It is a story of reconciliation that was happening long before reconciliation, as we know it today, was a "thing." It is a story that begins with curiosity, grows through relationships and understanding, and arrives at respect. And that is the essence of the path to reconciliation.

From Craig Macdonald

Hello. My name is Craig Macdonald, author of the *Historical Map of Temagami*. I came to this project in 1966 because of my love of canoe tripping, canoe sledding, and extended snowshoe travel. I had been guiding canoe trips for Kirk Wipper based out of Minis Kitigan on Lady Evelyn Lake, and I was friends with Teme-Augama Anishnabai Elder Peter Albany and his family, who lived at that time in two wall tents at Mowat's Landing on the Montreal River. Peter was very interested in encouraging and helping me record a priceless, undocumented Indigenous heritage of over twelve hundred portages, seventy-five snowshoe trails, and over six hundred Indigenous names for various geographical features and their meanings and associated stories. Peter was very concerned that this information soon would be lost forever.

This was the impetus that drove me to undertake this massive piece of work over the next twenty-six years, which ultimately resulted in the creation of the map.

From James Raffan

How I came to this project is a cute tale. At some point, long after we all knew that Craig Macdonald, a hero of Temagami, ought to have his life and work celebrated in a biography, the Grants — Richard and Victoria — approached Craig and asked if that was something he'd be comfortable with, and if a writer called James Raffan might be assigned the task. Apparently, he said, "Sure, but I doubt if James would take it on." The Grants then approached me and asked if I might consider writing Craig's biography. I said,

"Sure, but I doubt if Craig would be comfortable with me doing that." So it was a match made in a double balk.

I've known Craig, who was a staff member at the summer camp I attended, since I was about five years old. So he was instrumental in some of my early learning about canoe tripping and living comfortably in the woods. We've known each other for over sixty years. Over the years, I started travelling with him, mostly in the winter, on expeditions he organized to test some of the traditional travel techniques and technologies he was learning through his research with First Nation, Métis, and Inuit travellers. I have great respect and admiration for Craig and what he has done with his life, as I do for his wife, Doris, and their three children, who have made it possible for Craig to follow his passions to develop what I think is a truly astonishing body of work. To be clear, however, I researched and wrote *Echo Maker* under contract with the Temagami Community Foundation and, willingly, have no particular hold over what happens to the manuscript. In the spirit of doing things differently, in the wake of a country trying to respond to a pressing need to honour Indigenous knowledge and knowledge holders in new ways, I'm happy to follow their lead and trust that the published version of the story will be a book that will reflect well on everyone involved.

From Meghan Macdonald, Publisher at Dundurn Press

When people at Dundurn Press first started talking with Victoria and Richard Grant of the Temagami Community Foundation about publishing the book by James Raffan about Craig Macdonald and the map of n'Daki Menan and the *nastawgan*, we were honoured to help support it. It is such a remarkable story — indeed, we are hoping to mark it on paper — and we wanted to use our skills and knowledge to help create the book. I learned about it from my colleague Kathryn Lane, a quiet talker and a person who really helps me to listen to ideas that find their way to me in their own time, and from Victoria herself, who always makes me smile and think deeply about our conversations. It's hard to believe we've only had two video calls and an exchange of email as of the time I'm writing this. I was drawn to the story of Craig — who has walked the land, learned the language, and listened with an open heart to the people of the Teme-Augama Anishnabai community — and I wanted to find a way to come to the story with honesty and respect.

Dundurn Press has existed on the unceded territory of the Haudenosaunee, the Anishnaabeg, and the Wyandot Peoples for fifty years, and the stories we have helped tell have largely been Western settler stories that minimized or ignored the stories of the people of this land from time immemorial. And still there was a keen connection to the land that was part of each of these stories. The Québécois word for First Nations Peoples is *autochtone*, an ancient Greek word, meaning "of the same land," or "of the same place." Dundurn helped tell stories about the land in the past. We want to tell stories of the land in the future. I will make mistakes, I will fall down, and I will struggle to express myself in the language of Truth and Reconciliation. But I want to try.

I feel we must find a way to hold the stories of Dundurn's past together with the stories of a past we are still learning to recognize and care for. The story of Craig, told by James, and the knowledge kept by the Teme-Augama Anishnabai community and the Temagami Community Foundation is one such story. The depth and import of it is beautiful and breathtaking, and Dundurn is humbled to play a part, in partnership with the Temagami Community Foundation, to help learn about Craig Macdonald, the Echo Maker, the map of n'Daki Menan and the *nastawgan*, and the Teme-Augama Anishnabai People.

A Note Regarding the Sharing of Traditional Indigenous Knowledge

IT IS UNDERSTOOD THAT MUCH OF THE INFORMATION REFERENCED in this book and its companion works — the *Historical Map of Temagami* (Ontario Ministry of Natural Resources, 1996) and Craig Macdonald's "Traditional Sledding in North America" (Canadian Canoe Museum, 2025) — was gathered through personal relationships established by the author with Elders of the Teme-Augama Anishnabai and other First Nation, Métis, and Inuit communities throughout Canada, beginning in 1966. These informants (co-authors), listed at the back of this book, understood that Craig's prime intention in gathering the information was to preserve it in perpetuity. While this knowledge was willingly passed on to Craig for this purpose, he respectfully affirms that the knowledge itself, and the stories that surround it, belong to these informants and to their home communities, as captured latterly in the principles of ownership, control, access, and possession — more commonly known as OCAP® — that now inform the sharing of Traditional Indigenous Knowledge.

A note regarding Indigenous terminology: Because, with few exceptions, the information in this book was gathered orally, the spelling of terms drawn from these conversations is largely phonetic. In some cases, the author has inserted hyphens to aid with pronunciation or to indicate the etymology or origin of complex terms.

Introduction: Binaakwe Geesis — Ma-kõmin-ising, Te-mee-ay-gaming

WITH HIS CANOE LOADED AND HIS PARTNER, ANDREW COUCHIE, headed home for one last load of traps and a goodbye to the family, thirty-something Larry Turner took a tailor-made cigarette out of its cardboard packet and looked up at the full moon — Waw-wi-ay-zi Debik Ghesis, the nighttime sun. It sat bright and low in the west, as if it might soon be swallowed by the deep waters of Temeeaygaming. He reached forward and gently rolled the cigarette back and forth between his fingers. Shreds of tobacco, golden in the moonlight, fell into the silvered water at his feet.

In what was as much a habit as a prayer, he looked skyward and acknowledged Ojiig-anung, the fisher, and Maang-anung, the loon. In Maang's tail feathers was Giiwiidin Anang — the bright and strong North Star, the only one that never turns — which would guide them. In Larry's mind, stories flickered, times when his family first pointed out these constellations and taught him how they turn with the seasons and provide the nighttime reassurance that all is right with the world, that things are where and what they should be, and that it is always a good thing to give thanks at the beginning of a journey.

He turned away from the big lake and acknowledged Wabanung, the spirit keeper of the east. Strands of smoke from cabins on Makõminising (Bear Island) spoke of the days when whole families would have been loading

Andrew Couchie.

Larry Turner in later years.

up their *wiigwaas jiimaan*, their canoes made of the legendarily fine bark of the deep-water region, at this time of year to head for their winter trapping grounds. Makõminising had always been a place to gather in summer. But the arrival of a Hudson's Bay Company post in 1876 had started the change. Bear Island Day School was created in 1903. It was at first a summer seasonal school, but following government edicts, it then separated children from their parents in the winter too.

Larry thought about where each family was in n'Daki Menan, this land of ours. The Pauls, the McKenzies, and the Potts to the south and east. The Beckers and the Fridays to the east. The Pishabos to the northeast and the Katts to Nanwawkaming, which had never really been the same since the logging dam flooded them out. And then, to the west and north, the Misabis on Obabika, the Moores on Wawweeaygaming, the Twains beyond, and, of course, his own family, who had always trapped the western extent of n'Daki Menan. Everyone part of the Teme-Augama Anishnabai, the Deep Water People.

But on this day, Monday, October 20, 1938, while the families slept, the able men packed their kits in the shadows of Debik Ghesis and headed out, on their own or in pairs. They helped each other as they moved out from Bear Island along water and land trails — called *nastawgan* in the language — that were as old as time, to trap for family livelihood, for furs to trade and sell so that their young ones might grow up strong and healthy in a changing world.

More tobacco filtered through his fingers to the ground as Larry turned slowly toward Wawonong, the spirit keeper of the south. Rising before him against the deep indigo portent of coming dawn was Bugonagiizhig, a cluster of stars that were part of his very first memories of home. The seven daughters of the moon and sun. So recognizable, marking in memory the moon of falling leaves.

And following close behind was Biiboonkeonini, the main part of which was a constellation his summer fisher friends and clients at Temagami Lodge called Orion. He pulled another cigarette, special *asemaa* for the winter maker. May the ice be thick and strong. May the fur be thick and luxurious. May the season be safe, silent, and still from now until Biiboonkeonini was less prominent in the sky and Mishipeshu, the great horned lynx, was overhead. If Mishipeshu summoned an early thaw, slush and thin ice could complicate the end of the best trapping time of the year. Although Mishipeshu was not visible in this night's sky, Larry offered some special shred of *asemaa* to honour the keeper of the underworld.

Larry looked at Bugonagiizhig. In his memory, he heard the voice of his *nokomis*, his grandmother, telling stories by the reassuring crackle of a wood stove. "There were seven children," she said,

> who loved to dance and play rather than help out their parents around the camp. The children's mother went to the elder to seek advice on this problem and was told to put stones on their food. Somehow, it was hoped that the children would appreciate the value of hard work if they were forced to remove big stones from their food before they could eat it. Unfortunately, the plan backfired. And one day, fuelled up with the food they had pulled out from beneath the stones, they danced so hard that they danced up into the sky where they can be seen to this day, especially in fall and winter, when families were in camps across n'Daki Menan.[1]

But there is a darker side to Bugonagiizhig. Grown-ups would speak about this in hushed tones when Larry was a boy. Bugonagiizhig, the Hole in the Sky. Seven clustered stars, visible even with the light of Waw-wi-ay-zi Debik Ghesis. The opening between the Earth and the star world, a mysterious portal where great spirits collide, where great reckonings take place, where there are winners and losers and fiery encounters of many kinds. No mention of the seven daughters of the moon and the sun in those breaths. Same stars, different story. Respect always.

Footsteps and the jangle of steel traps approaching from up the hill broke Larry's reverie of thanksgiving, his trip around the universe in a moment by the lake. Larry turned to see Andrew, lit by the setting moon from the lakeside, silhouetted against the brightening arch of ochre in the east as the stars faded overhead.

They had spent a couple of weeks preparing: pulling supplies together; cleaning tarps, tents, tin stoves, cook sets, traps, and snowshoes; getting their winter clothes organized; and making seaworthy their aging eighteen-foot Peterborough canvas-covered cedar canoes. Each man's kit was shaped by his preferences and predilections. Flour: two hundred-pound sacks. Lard: three twenty-pound pails. Other supplies were stored in various grimy hand-sewn ditty bags and green canvas packs with leather tumplines: five pounds of salt, twenty-five pounds of brown sugar, ten pounds of macaroni, ten pounds of rice, six packs of matches in screw-top tobacco tins, one pound of dehydrated onions, half a pound of black pepper, ten pounds of baking powder, half a pound of baking soda, two ten-pound slabs of smoked bacon, twenty pounds of dried beans, ten pounds of salt pork. Tent and stove. Sleeping rolls with HBC blankets and one traditional rabbit skin blanket to cover them both. Axe and saw. Belt knives and a favourite sharpening stone. A mix of traps: Newhouse 3s, Oneida-Victor 4s, and lots of snare wire, poor man's traps. All pretty predictable. The mystery item in Larry's kit on this run was a suitcase radio and two big, heavy Delco glass batteries filled with sulphuric acid. Through experience he knew that if the batteries were broken or tipped, the liquid would likely seep through the floor planking and burn through the canvas covering of the canoe.

It was all here, tucked together and ready for the annual fall move to the northwest.

And not a map to be found anywhere — except the ones inside their heads.

The dawning day was as clear as the recollection of this time last year, and the year before that, and so on deep into memory. Leaving Makõminising was less a departure than a continuation of motion, a dance within a universe of places stitched together on a broadcloth of legends, stories, and experiences that drew from the deepest depths of Temeeaygaming and extended to the farthest reaches of the stars and every conceivable horizon revealed in all directions of the medicine wheel. Connecting the traveller, the listener, the teller, the past, the present, the future.

And connecting the two men on this journey to all of that was the canoe — the *jiimaan*. Her rhythms had been written into their bodies, minds, and spirits in the womb from the moment of conception, maybe before. As toddlers they rode on the load as the family relocated from summer quarters to winter trapping grounds, moving forward to the space in front of the bow seat when they needed a stretch, or moving back to pester the parent in the stern. Always, no matter what the condition of the waters, whether calm or whipped up by winds from whatever direction, balance was the key, lest they be berated by the grown-ups.

Finding the natural balance of the *jiimaan* in the water was the only acceptable way to move. Adjusting to the side-to-side oscillations; the gentle surges as the paddles pulled; the movements of others, each in their own time. Being in the canoe, buoyed by the waters, connected to the heavens, breathing the stories written into its skin, was not so much learned, like a school lesson, as it flowed into, through, and around young Anishinabeg with the passing of the hours on the water with family and the energy of the ages that swirled.

It might be a voice, a colour, a sound, a smell, a particular location or scene that triggered a memory. There was the realization that, in addition to language and family, the *jiimaan* was the physical connection back to time immemorial and to the beginnings of the Anishinaabe People. Like the people themselves, the *jiimaan* was a gift of the Creator: its skin protected everything inside; its cedar ribs and planking were the bones and muscles that gave it form and character; its spruce root lacings stitched the *wiigwaas* together in those first *jiimaan*; and, of course, the pitch waterproofed those seams in the original canoes, holding everything together, especially the spirit of the vessel that made paddling a spiritual practice. To paddle was to praise all creation, without saying a word.

After a quick scan of the surrounding area for anything they might have missed, the two trappers gently released their canoes from the shore and then stepped aboard. Customarily, with just one paddler in each, the trappers would have left a gap around the stern seat to squeeze into or, if there wasn't enough space for their feet there, they would perch on the stern deck and paddle from there. As the sun started to peek over the trees of Makōminising, they took their first strokes. When they cleared the community wharf, they swung north on calm water.

The canoes were loaded to the gunwales with food and supplies. Over a thousand pounds of cargo and men, but well within the load capacity of these sturdy eighteen-footers. With the weight evenly distributed, it didn't take much urging to get the boats going, and they tracked well in the water. With the trappers sitting as far away from the midpoint of the canoes as they could get, it took only a suggestion of outward force with the paddle to keep straight in the water. Sometimes they ruddered at the end of a forward stroke to steer, and sometimes, just to keep the body limber, they switched sides and pulled away to stay the course with a minimum of effort. The trappers each had their own paddling cadence and rhythm. For the most part they stuck close together, within earshot, though little was said as they kept an eye on the sky and the wind.

By noon, they had cleared the islands west of Makõminising, crossed the most exposed part of Temeeaygaming, passed the beach at Kawnaydownkawg (Sand Point), had a smoke and left a sprinkling of tobacco off the pictographs just before Wigwasobawshing (Birch Narrows), and were heading toward the place at the head of the northwest arm they knew by the name Kitchaywaskigamawg. Here they made a nearly right-angled turn to the west that led them to the portages into Obabika.

Just before leaving the open waters of the big lake, they heard in the distance a familiar murmur that slowly morphed into a distinctive chugging sound. Almost in unison, they set down their paddles and turned to wave to Captain Bill Reynolds as the aging steamer *Belle of Temagami* slowed for a quick stop at Camp Bigiwe before swinging around to the north arm on its daily run to Camps Keewaydin and Wanapitei. This late into the fall, most of the summer activity on the lake had ceased, but the caretakers were still battening down the hatches for winter. So *Belle*, part of a family of passenger vessels on Temeeaygaming, continued to run people and supplies until freeze-up, as she had done since entrepreneur and innkeeper Dan O'Connor launched her in 1906. Camps and hotels brought wage work and summer employment to both Larry and Andrew, and to almost anybody else on Bear Island who was interested. But life on the lake and across n'Daki Menan had never been the same, particularly after the Temiskaming and Northern Ontario Railway line reached the town of Temagami in 1905.

Today, over the gentle knocks of their hand-carved paddles on the gunwales and the insistent chattering of overhead geese, the low howl of a distant train whistle came to their ears on a freshening southeast wind. The taste of the wind confirmed what the ring around the moon and the ground-swirling smoke from the home fires had already told them. Rain was on its way.

In midafternoon they pressed on through Waskiggamawsing (Bent Lake, also known as Obabika Inlet). Here they had to parse the jumble of food and supplies they'd loaded by the light of the stars and the moon into carryable loads so they could tackle Neebinee Oniguming. This would be the first and perhaps the shortest of many portages on the twelve-day journey to Larry's trapping camp on an island in Sawgidjeewayawgamawk (Solace Lake). In other years, members of the Misabi family from North Obabika might have been travelling along at the same time, or perhaps some of the Moores and maybe relatives, like Andrew's father-in-law, Pete Commanda, and they would all have pitched in to do the carry. They would have made sure to keep track of whose kit was whose, especially when it came to stacking it up on the other end of the portage, ready for reloading for the short paddle to the next carry, which would take them through to Obabika. This day there were just the two of them. Together they set about tumping and humping a ton of canoes and gear overland.

Saying little, they veered right through the small pond that led to the trail, Abikonabikush Oniguming, and unloaded onto its bouldery threshold as the daylight started to wane. The patched canoes, which had soaked up a little water throughout the day, weighed perhaps a hundred pounds each. It was possible for the paddles and a leather tumpline to be lashed to the middle thwart of a canoe so that one man could carry it, but they opted to lash the paddles to the stern-quarter thwart and carry it together, one in between the lashed paddles and the other guiding the way with the bow on his shoulder, each with a smaller load strapped to his back. Experience had taught them that every little efficiency saved energy and time. On this rocky portage, it was good to have one man at the front of the canoe picking the best footing for the man in the rear.

It was almost dark by the time they'd finished three trips back and forth on Abikonabikush. They left a couple of loads for the morning. It was not a trail to walk in low light — too many little jogs, roots, and rocks. They ignored the massive concrete fireplace the Ontario Forestry Branch had built for the tourists to cook their fish in during the summer months. Instead, they arranged rocks and twigs between the two overturned canoes and built a small fire to warm an iron pot of beans they had carried from home and a pail of lake water for tea. With chunks of bannock broken from rounds they had cooked and packed in a canvas bag for the trip, they ate in silence and listened as two young loons cruised by against the day's last glow in the western sky. Before turning in, while Larry had a smoke and finished his mug of tea, Andrew walked the shallow edge of the bay, his senses tuned for spawning fish, movement, tracks, prints, stories written into the minutiae of the land, or perhaps idly scrolling through memories of other days, other nights at this very place.

Perhaps the fancy concrete fireplace brought happy recollections of shore lunches when they'd worked as guides for the Temeeaygaming Lodges in the summer, but it might also have reminded them of the mixed blessings that the incursions of white southerners had brought to the people of n'Daki Menan. Sounds that occasionally floated their way on the breeze — piercing steam whistles, the chug of the mine engines, and the rumble bulk ore movements at Golden Rose Mine on Kawwawkonikkawgamawk (Emerald Lake), a few miles south of their campsite — these were all sounds of change.

Like the arrival of the Hudson's Bay Company post on Makõminising in their great-grandparents' time, these social and technological additions to their lives had brought benefits, yes, but they also visited upon the Teme-Augama Anishnabai a gradual erosion of independence and control over their traditional lands. The concrete fireplace the two trappers had chosen not to use was erected by the rangers of the Ontario Forestry Branch to diminish the risk of fire in the Temagami Forest Reserve.

The creation of this reserve, which essentially annexed the full extent of n'Daki Menan, from the Montreal River in the west to the Sturgeon River and beyond in the east and to the Mattagami River valley in the north, imposed strictures on land use practices, like trapping. Indeed, beaver trapping had been prohibited within the forest reserve in 1928 and was reopened only in 1937 by the Ontario government. The reserve also sought to protect n'Daki Menan's great stands of virgin red and white pine trees through rules that precluded those who lived there from cutting wood in many places, even for homes or heating. That concrete fireplace, with its litter of burned cans and fishbones, was a monument to all of that.

The store and the school on Makõminising and the proximity of the railway in the growing town of Temagami had slowly pulled families off their traditional territories throughout n'Daki Menan in the winter. Besides this, the most significant impact of these changes, particularly the railway, was the arrival of white trappers — or would-be trappers, like the illustrious Englishman Archie Belaney, or Grey Owl, who had paddled up from Mattawa into Temagami with Bill Guppy back in 1906. The trappers came north mostly by train, with bush skills, perhaps, but with no real authority to "share" in the fur harvest of n'Daki Menan. In times before the changes, families would leave portions of their traditional territories fallow, rotating every few years to allow the fur-bearing populations to recover. However, the he-with-the-most-furs-wins ethic of the white trappers did much to erode this sustainable natural stewardship of the fur harvest that had allowed the Teme-Augama Anishnabai to live in harmony with the lands and waters of n'Daki Menan for millennia.

It would be another few years before government registry of traplines was established, and it was in the midst of all these changes that Larry Turner and Andrew Couchie were making their way to their chosen trapping grounds. They were perhaps among the last Teme-Augama Anishnabai with language and deep land knowledge of winter and summer trails and place names, of the lore, stories, techniques, and ethics of the fathers and grandfathers, and their fathers and grandfathers before that.

• • •

In the irony of ironies, in September of 1929 the Burden Expedition came to Temagami to make a Hollywood film, called *The Silent Enemy*, about the plight of the Ojibway. From the point of view of Hollywood, the Ojibway's "silent enemy" was hunger. Although more than a hundred Bear Islanders were involved in the production, only six of whom had ever seen a moving picture, the film starred Sylvester Clark Long, also known as Long Lance, a Black man with not a drop of Anishinaabe blood. While this was one of the very last silent movies ever produced, one of the other stars, a Rosebud Sioux man from South Dakota called Chauncey Yellow Robe, introduced the film with a short piece of sound-synched narration. Here's what he said:

> This picture is the story of my people. In the beginning, the Great Spirit gives us these lands. The wild game was ours. We were happy when game was plenty. In the years of famine, we suffered. Soon we will be gone. Your civilization will destroy us. But by your magic we will live forever. We thank the whites who helped us to make this picture. They came to our forest. They shared our hardships.... That is why in this picture look not upon us as actors. We are living our own lives. To us, we live as yesterday. Everything you see is as it always has been. Buckskin clothes. Our birchbark canoes. Our wigwams. And our bows and arrows. All were made by my people, just as they always have done.... Many of them are still in the forest, hunting the game that is ever growing less. Still living the great drama of the north. They struggle for meat, a never-ending fight against the silent enemy.[2]

And with that, the image of Yellow Robe fades and the word "HUNGER," in full capital letters, scrolls in white against a black background.

The fictional story, written by W. Douglas Burden, Richard Carver, and Julian Johnson, is seriously hokey, as are the various non-native wild animals and far-fetched scenarios that are supposed to represent everyday life in n'Daki Menan. But the film did garner some success. In fact, in February of 1938, the same year that Larry Turner and Andrew Couchie headed out, Bear Island entrepreneur John Turner and his neighbour Madeline Katt Theriault travelled to the New York and Boston Sportsman Shows, on the strength of contacts established through the making of *The Silent Enemy*, to demonstrate Anishinaabe crafts and customs. Madeline brought her nine-month-old daughter, Virginia, in a papoose cradle. Her husband, Alex Mathias, stayed back at Bear Island to tend their other daughter, Dorothy, who was seven years old at the time.

That August, inspired by the interest of audiences at the sportsman shows, Madeline was able to convince a number of her friends and neighbours to help create a similar dramatic tableau at Bear Island for that year's summer tourists. In her book, *Moose to Moccasins*, she provides a sketch of how this all worked: "Our setting consisted of one birchbark tepee, one brush lean-to, one brush wigwam, four baby hammocks and one open fire. In the performance, everyone who took part was fully dressed in costume. Chief Pishabo, wearing feather head-dress, took care of the fees. He was supplied with a moose leather bag into which people dropped money, whatever amount they desired to give. Eva McKenzie, his little helper, rewarded the money-givers with a piece of moose leather as a souvenir to attach to their clothing."[3]

History doesn't record if Madeline's husband was among those gearing up and heading north to the family trapping grounds in October of 1938. He was well into the early stages of a battle with tuberculosis, to which he succumbed, like too many Teme-Augama Anishnabai, in 1940.

• • •

So it was with all this context swirling that Larry Turner and Andrew Couchie awoke under their canoes beside the concrete fireplace at the Obabika end of Abikonabikush Oniguming. Under an overcast sky, with the *zhaawani-noodin* (south wind) rising, they loaded their canoes quickly and immediately paddled out around the point on the east shore. They crossed the lake in a southwesterly direction, knowing that as the sun heated the day, the opposite shore would provide shelter should the wind increase in intensity.

Their loaded long canoes tracked easily on a quartering wind, crossing a scurry of invisible lines marking other dreams, other destinations up and down this storied lake — the continuity of Anishinaabe habit — since time immemorial. The hypnotic pull and dip of a paddle on this route always brought memory flashes of other days, other journeys. Images of ancient Iroquois raids, spawning fish, willing moose at water's edge, children giggling in loaded canoes, spring sugaring. Of the living rocks that whispered through the waters and white pines of Jee-shi-kun Abikong on Misabi family territory at the north end, and of the cold campfires of curious tourists and white loggers throughout.

How reassuring it must have been for these two travellers to reach the rocky western margin of southern Obabika and to rest beside rock images, etched in red ochre by their ancestors, that confirmed the connections to the heavens and the centrality of the canoe in the Anishinaabe way. The tally marks, the work of many hands, spoke of a relationship with the land, the water, the sky, the spirits of the place that had sustained generation upon generation of their forebears. These shining rocks, all along this shore, had been part of every journey these two men had ever known. And as their parents did, and their parents before them, they made an offering of copper coins or tobacco, thanking the rocks for their enduring wisdom and asking for a bountiful season and safe passage to where they were going.

They swung west around Kawgawgeewabeekong (Raven Rock) and on through the narrowing west bay of Obabika to the familiar portage to Wawweeaygaming. They couldn't walk this worn trail without thinking about its story.[4] They travelled as much on a storyscape as on a physical landscape or on a course set out for them in the heavens. Sometimes the story was triggered by the place names or by the position of a particular star. Sometimes the story just came with the experience of being in a particular place or season. Some places and the stories they held should be celebrated and spoken about openly. Others not so much, for a host of private and personal reasons.

The clue about what happened here came from Cheench Point on Wawweeaygaming, where it is said Kanachinz Whitebear was buried sometime around the 1860s. How Kanachinz lost his life is attached to the portage and to an altercation with his brother, Shakosay, and his brother's wife, Managit — tension resulting from family affections gone wrong, as Kanachinz desired a relationship with their daughter, his niece. Larry and Andrew would likely have heard multiple versions of this story throughout their lives. Some would have related an axe and a knife as murder weapons wielded by one or the other of the perpetrators; some might have assigned blame for the fatal blow to Shakosay, and others to his wife, who became "Stabbing Woman" in the evolving narrative.

As the two trappers completed the haul of all their gear across the storied portage into Wawweeaygaming, their attention might have been drawn by the sounds of dogs barking or children's voices from Charlie and Alice Moore's place on the north side of the lake. It's hard to say which of the Moores and their children were on their family territory that October day in 1938. The three school-age girls — Laura, about fourteen; Eva, twelve; and Rita (Bubsy), ten — might have been back at Bear Island with their mother, along with brother Sonny, who was nine, and baby Emma Doris Elizabeth, who was still breastfeeding. But Larry and Andrew would have stopped in regardless, especially with a long rain coming on as they crossed the lake. If Charlie was anywhere near there, they would have had a good chat about the finned, feathered, and furred comings and goings in this corner of n'Daki Menan as they enjoyed a feed of fresh whitefish or trout, which would have filled their bellies during the visit and their travelling pot for meals ahead on the trail.

For many Teme-Augama Anishnabai, Wawweeaygaming represented the western edge of the home "neighbourhood" surrounding Makõminising, where everyone travelled with ease in winter and in summer, knowing the routes, the portages, the places to stop, the possible locations of fish and game, the spiritual sites, and the stories that stitched all of this together in collective memory. From here, the solidarity of the routes, place names, histories, and conditions would have been of equal value in the mental maps of the people like Larry and Andrew and the families like the Moores for whom this corner of n'Daki Menan was home.

With goodbyes and good wishes for the trapping season coming up, the lads would have re-embarked for the southwest corner of Wawweeaygaming to drag their loads through the low autumnal waters, sloughs, push-ups, and beaver dams of Wawzushko Zibi (Muskrat River). Between wading and paddling and dragging their canoes, they attended to every nuance, every sign of the condition of the fur-bearers in this environment. The size of the beaver chew marks on the trees would tell them about the age pyramids of the beaver families in the area. The size and nature of the muskrat push-ups would reveal clues about their success as a species in the months since the two trappers had been through here. It being rutting season, they may even have encountered a bold male moose, whom they would have thanked for making himself known to them. But very likely they would not have hunted the moose for the time and effort it would have taken to dress the carcass and then to dry or move the meat and the hide before they spoiled, even if they had room in their canoes for another few hundred pounds of cargo, which they most certainly did not!

Wawzhushko Zibi took them to Obawbika Zibi and then twisting down through oxbows to the carry around the impassable waters at the confluence of the Obawbika Zibi and the bigger, higher-volume, Nahmay Zibi (Sturgeon River). For years, Andrew had kept a camp on the island just downstream of this confluence from which he'd trapped. Over time he slowly moved north, forced by diminishing fur reserves and the activities of J.R. Booth's son, who had crews in the area actively logging and gratuitously mucking with shorelines and water levels from time to time. Depending on progress as they wound their way down from Wawweeaygaming, they may have stopped at Andrew's island camp. The amenities there and their intimate familiarity with the place made it an easy spot to spend the night, with a minimum of unpacking, and allowed a quicker start north in the morning.

Right about here, the character of Nahmay Zibi changes as it moves upstream, from a more placid river flowing through the sandy soils of the Obawbika Mushkooding (Obabika Prairie) to a more aggressive pool-and-drop riparian passage through the hard rock of the Canadian Shield. Heading upriver from Andrew's old camp past Upper Goose Falls, the place they call Amik Gabuddow (Beaver Walkover Trail), the two of them used their knowledge of curves and current — even in the low waters of October — to paddle upstream to the cabins at the old settlement of Abondiako and a little ways beyond that. At that point, the nature of the river was such that the only sensible way to get up into the high country beyond was to commit to a long and ugly four-mile ancient *nastawgan* linking Nahmay Zibi and the waters of Kawcheekawbkiggaming ("big side rock," now Yorston Lake).

It took the two trappers three days to negotiate this difficult but important connection in one-mile chunks, camping right on the trail, as they had done before, as their ancestors had done before that. The alternating northward trudges with heavy loads and unfettered return walks provided two very different mindsets to engage with the tasks at hand, to voyage in the inner spaces of memory and imagination. On the southward wanders, with nothing pulling on their foreheads, shoulders, or backs, they had a chance to see the claw marks of bears climbing poplar trees for spring leaves and square blazes on pines where their forebears would have gathered pitch to repair bark canoes damaged in transit. They saw the progress of moss and lichen on the south side of saplings and noted all of the other minutiae of nature on which a successful Indigenous trapper's existence depends.

Seeing a particularly unwary spruce grouse on successive trips up and down the trail might have prompted one of them to break the monotony by chopping out a sapling pole and affixing a little snare-wire loop to its end. Carefully approaching the bird head-on (the only way) would have yielded a couple of plump grouse breasts to roast on a stick — *abondiako* — with their evening tea.

Fellow trappers working near Ojeeg Pukwudinŏng (Fisher Mountain) — whom Larry and Andrew would surely see on their way back to town in December, if not before — might have carried on up Kawcheekawbkiggaming and into the river that feeds it from the north. But Larry and Andrew turned west through Misandip (Bull Beaver) Lake, following the Pigeon River to Onegoshee Gaygodayg. Many Indigenous place names refer to characteristics of the topography or the nature of the feature itself, while other names connect places and features to people, through story, or to events, some of which have humorous overtones to them. Such is the case with Onegoshee Gaygodayg (Hanging Shit Place). If anyone asks, sniggering, what on earth this name means, it refers to a time when someone dressed a moose here and, for reasons not revealed by the name, hung its intestines in a tree that for years marked that exact spot.

Eventually, after twelve days of hard labour and joint effort — which flowed through their lives as just something they did to carry on carrying on — Larry and Andrew reached Sawgidjeewayawgamawk. They arrived at this splendidly scenic spot in n'Daki Menan's high country on a portage from the south. Sawgidjeewayawgamawk means "going over a hill where you can get a view over everything." So much more apt and expressive than Solace Lake, the name those who came later afforded this place. This lyrical and expressive toponym was inspired by a point farther up the lake, arrived at from the east. Larry and Andrew would take this portage, their first on snowshoes (and, hopefully, toboggans laden with furs), as they travelled back to their families in December.

Together they made their way up to the crossroads at the centre of Sawgidjeewayawgamawk and to Larry's established camp, on an island between the largest island on the lake and the western mainland. From here, Andrew would move to Regan Lake, five miles north. But now it was time for Larry to see how his site had fared since spring and to set up the tent that would be his home for the next couple of months, and for the two of them to do what he'd been thinking about every time he had picked up those two heavy and awkward glass batteries and the suitcase radio they had lugged all the way from Makõminising. While Andrew took a coil of snare wire and ran an antenna line up a tree just outside of the canvas-wall tent, Larry set up the radio on a low table he'd fashioned a previous year and left to weather while he'd been gone. He then attached the batteries. And as darkness fell, with the last of the Moores' trout heating on the tin stove, Larry attached the antenna, turned on the radio, and began spinning the dial to see what they might get.

Radios were all the rage in Temagami in 1938, and most households with a little money to spare had one. But that was on Bear Island, where the batteries were more or less stationary and not subject to breakage of the lead electrodes or spills of the corrosive acid that the glass cells contained. And although called "portable" for home use, the radios of the day, like the Stromberg-Carlson radio, were anything but. Nevertheless, Larry had done well with his trapping in the previous few years, and he had the wherewithal to add this kind of communication capability to his kit. He knew how lonely it could get at his camp on Sawgidjeewayawgamawk between rounds to check traps.

These were AM radios, broadcasting with waves that, unlike line-of-sight walkie-talkies, would bounce off upper layers in the atmosphere. There were just a handful of stations with powerful AM transmitters in the 1930s, all

Saw-gid-jee-way-aw-ga-mawk
War of the Worlds
Waw-wee-ay ga-ming

Chay S.
O-bawb-ika S.
Ko-ko-mis Waw-bee Kōng
Sho-mis Waw-bee-kōng
Chee-skon-abikōng S.
Kaw-kin-aw-ga-mawk
Kaw-bi-kway-beega S.
Kaw-minisi-wōng S.
Kaw-mahnu-zhi-mawk
Kaw-gaw-shab-koka W.
Waw-wash-keshing M.
See-ak-say S.
Chew-wun-ay S.
A-chit-a-moo S.
Kaw-wawm-inash-ing W.
Mush-koday Pi-jeeki Pa.
Ogas W.
Kawb-jee-way-nee-wōng S.
Man-i-doo W.
Ko-ko-ko S.
Man-i-doo M.
Man-i-doo Nas-suy-bee Pa
Kaw-chee-kawb-kig-ga-ming S.
Ko-ko-mis W.
A-chit-a-moo N.
Kaw-waw-anda W.
Chay-o-bawb-i-kōng
Kaw-o-gonse-i-wōnk S.
Meeda-wawp-nish S.
Was-kig-ga-maw-sing
Kit-chay Was-kig-ga-mawg
Mo Agodayg W.
Waw-bee-zhaysh Wawbinee-Kaw-ning W.
Kaw-waw-bish-kaw-ga-mawk
O-bawb-ika S.
Pick-wawk-wido S.
Pishew Kee-ko-bee kwash-ko-nay S.
Kaw-nay-down-kawg N.
Kaw-miti-wawb-koka S.
Miskwa Wigwas S.
A-shig-gin S.
Kaw-waw-bish-kawg-ga-mawk
Ma-kōmin-ising
Anokiwin S.
O-gas-a-damsing S.
Kaw-kinō-zhay S.
Kaw-mawng-gonse-waw S.
Ko-ko-ko W.
O-gas-a-damsing W.
Kosh-kok-way-aw-ga-mawk
Machanis M.
Way-kwa O-bawb-ika S.
Saw-gi-hay-gōnse-ing S.
Pee-das Moosay S.
Guy-awshk-go-senda S.
Kay-tay Te-mee-ay-ga-maw M.
Waw-saks-sin-aw-ga-maw W.
Te-mee-ay-gaming
Kinay-oo
Shee-kawg S.
OUTGOING ROUTE (WEST)
Canoe route (all seasons)
RETURN ROUTE (EAST)
Snowshoe route (winter only)
OUTGOING AND RETURN ROUTE
Canoe route (all seasons)

in the United States: one in Chicago, Illinois; one in Nashville, Tennessee, which beamed out hugely popular programming from the Grand Ole Opry; and, from New York City, the Columbia Broadcasting System radio network, which blasted a signal to the world every night for anyone to hear.

As it happened, Larry and Andrew arrived at the camp on October 30, the day before All Hallows' Eve. And as Larry spun the dial, the strongest signal was from the CBS radio network. Over the miles and into that white canvas-wall tent lit with a dim red glow from the tin wood stove came a crackly voice: "The Columbia Broadcasting System and its affiliated stations present Orson Welles and the Mercury Theatre on the Air in *The War of the Worlds* by H.G. Wells." Their first language, and they were of the last generation of Teme-Augama Anishnabai for whom this would be the case, was Anishinaabemowin. This broadcast was in English, a language they had been compelled to learn to ensure good pay and tips in the summer, guiding southern clients for the owners of the Temagami tourist lodges.

After a musical sting from a live orchestra set the tone of the program, Orson Welles's commanding voice led them into the story:

> We know now that in the early years of the twentieth century, this world was being watched closely by intelligences greater than man, yet as mortal as his own.
>
> We know now that as human beings busied themselves about their various concerns, they were scrutinized and studied — perhaps almost as narrowly as a man with a microscope might scrutinize the transient creatures that would swarm and multiply in a drop of water.
>
> With infinite complacency, people went to and fro over the earth about their little affairs, serene in the assurance in their dominion over this small, spinning fragment of solar driftwood which by chance or design man has inherited out of the dark mystery of time and space.
>
> Yet across an immense, etherial gulf, minds that are to our minds as ours are to the beasts in the jungle — intellects, vast, cool and unsympathetic — regarded this

> Earth with envious eyes and slowly and surely drew their plans against us. In the thirty-ninth year of the twentieth century came the great disillusionment.[5]

Across the airwaves floated the now infamous episodic news reports of a meteor strike in the New Jersey countryside and the very realistic, breathless news reports of mysterious creatures with terrifying war machines and thick clouds of poison gas overtaking Manhattan. At some point in the middle of this debacle, Larry's fancy bush radio failed, so they never did hear how things ended. But they did hear enough to think that if all this *was* real, the scourge might well have reached their families back at Makõminising. Truth be told, there wasn't much they could do about it if any of this was true. They had slogged over many days to get to this place. Andrew's daughter, Cathy, later recalled that, in her father's telling of this story, he and Larry looked at each other in the moments following the interruption of the broadcast and shrugged, saying, "Well, maybe we'll be the last two people on earth."

In truth, the war-of-the-worlds motif was anything but new to these Anishinaabe men. The Nimkee-Binesi — thunderbird — on the Obabika rocks had reminded them of stories about great conflicts between worlds, stories that were very much a part of who the Anishinaabe People are. And, with equal verisimilitude, every sign of influence on their lives driven by southern appetites — Royal Proclamation, the arrival of the Hudson's Bay Company, the building of the railway, the creation of the Temagami Forest Reserve, schools, language imperatives, and all of the little things that governments had done to erode their way of life — spoke clearly of slow-motion cultural collisions in real time that could easily be understood as a war of the worlds happening throughout n'Daki Menan, right before their eyes.

In mid-December they packed up their fur bounty from two intensive months of trapping. They packed up their camps and stored the remaining supplies for the spring, when they would arrive again for spring ratting and return by canoe to Bear Island. Then they fastened their snowshoes, set the tumps of their toboggans across their chests, and crossed Sawgidjeewayawgamawk, heading east over the eponymous portage, now serving as the *bonkanah*, or winter trail, leading them home to Bear Island.

Meeting up with others at Kawchingwawkōng Lake, they camped, and at dawn the next morning, they climbed Ojeeg Mountain to its crest. In the distance, they saw just a hint of winter morning mist over Obabika, fourteen miles away, and over Temeeaygaming, indicating that although the waterways in the high ground where they had been trapping were well frozen, the two big lakes were still open. They'd have to wait a few days longer until there was sufficient ice to get them home to see, once and for all, if there was indeed a home still there. That they did. And that there was, a community just as they had left it.

They were home for Manitdoo-Giizisoons (Little Spirit Moon — Christmas at St. Ursula's Church at Makõminising) and a celebration of the start of the sun's slow return and longer days. Home with furs to sell. Home with the last vestiges of a way of life and a lexicon of place names, language, and belief that would fade as their children and grandchildren made other plans.

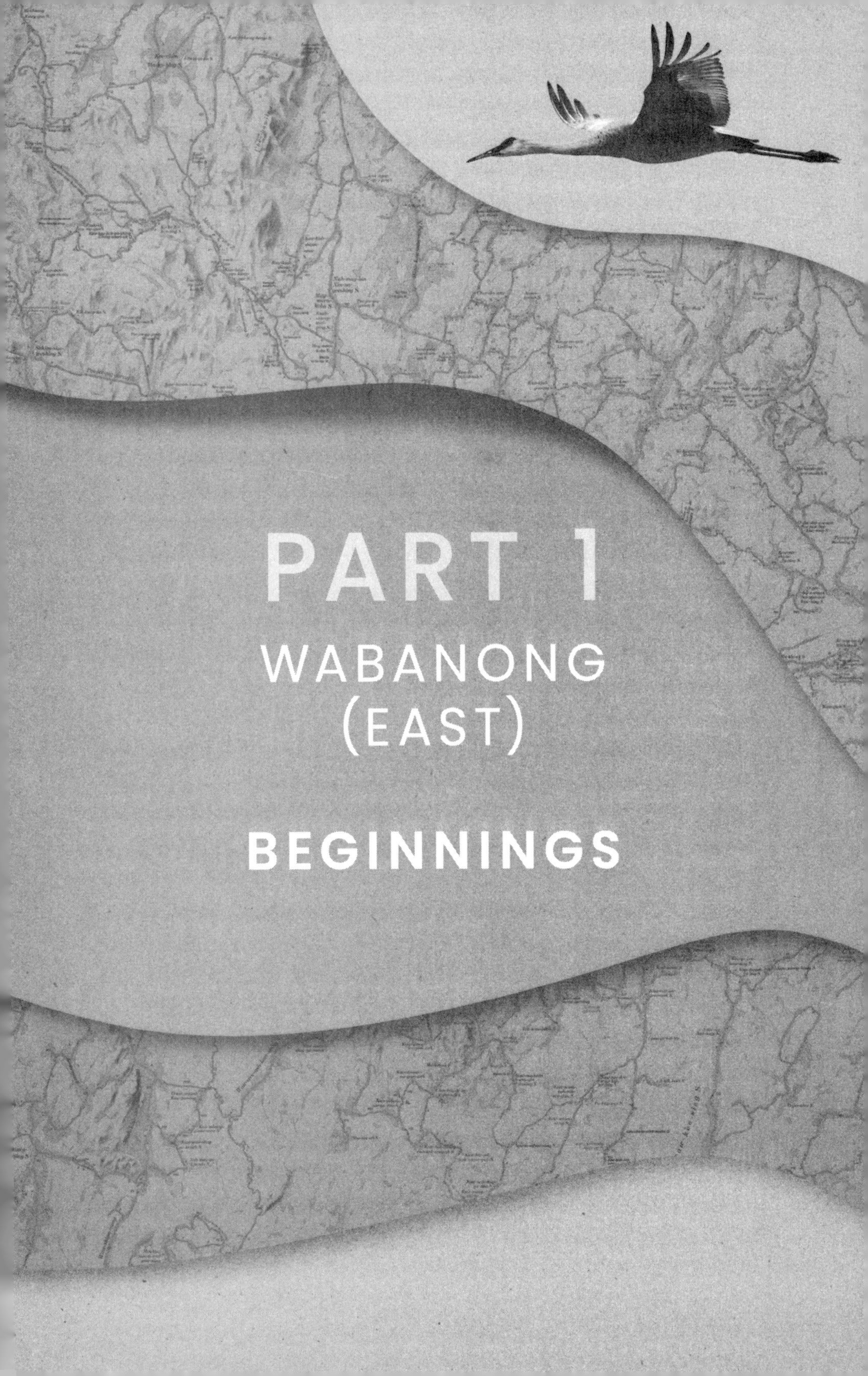

PART 1
WABANONG (EAST)

BEGINNINGS

Mr. Macdonald Testifies in the Supreme Court of Ontario

AT 10:00 A.M. ON THURSDAY, JANUARY 27, 1983, IN THE SUPREME Court of Ontario in Toronto, everyone stands and falls silent as the Honourable Justice Donald Steele takes his place on the bench in the matter of *Province of Ontario v. The Bear Island Foundation and Gary Potts, William Twain and Maurice McKenzie on behalf of themselves and on behalf of all other members of the Teme-agama Anishnabay and Temagami Band of Indians regarding their land claim*. Justice Steele looks across to the defendants' table to his right and addresses Bruce Clark, lead counsel for the defendants:

His Lordship: Yes, Mr. Clark, your next witness.
Mr. Clark: Craig Macdonald, please take the stand. My lord, this witness is just recovering from the flu, would it be all right if he sat down while he gave his evidence?
His Lordship: Certainly.

Craig Macdonald, thirty-six, is sworn in and takes the stand. He has in his hand a set of written notes and, unlike the lawyers, looks decidedly uncomfortable in a store-bought suit.

Q. Mr. Macdonald, I will move quickly through your qualifications insofar as they pertain to the evidence you are about to give. You were born in 1946 at Markham, Ontario and received your M.A. in biology from California State University in 1972, and an Honour's Bachelor of Science degree in biology from the University of Guelph in 1970; is that correct?

A. That is correct.

Q. With respect to your employment, since graduation, you have been employed by the Ministry of Natural Resources, Province of Ontario, is that correct?

A. That is correct.

Q. In the years 1972-73 you were a full-time specialist charged with the responsibility for canoe route research for Southern Ontario, located at Richmond Hill Ontario MNR Regional Office. 1973-75, full-time specialist charged with responsibility for research, planning and policy development for the recreational trail system of Ontario. This included canoe routes, and hiking, snowmobile, snowshoe, cross-country ski, bicycle and equestrian trails for Ontario.... 1975-77, charged with responsibility to study in detail the canoe routes and trails of the MNR Algonquin Region representing approximately the northern third of Southern Ontario.... 1977 to the present, stationed at the Leslie M. Frost Centre, Dorset, Ontario. Serving as the recreation specialist managing the recreation facility on the 58,000 acre property (canoe routes, access points, interior campsites, snowmobile and cross-country ski trails, et cetera).... Is that an accurate summary of your employment history?

A. Yes, it is....

Q. You have spent every summer as a youth canoe tripping. Throughout high school and university you financed your education by working as a canoeing guide in the bushlands of Ontario and Quebec. You have served as a guide for the Lake Superior segment of the Ontario Camping Associations' trans-Canada canoe pageant in 1967. Through the late 1960's and early 1970's you were canoeing around one thousand miles each summer. You have guided senior government officials and politicians in MNR related business. You have also guided on snowshoe trips in the northern Haliburton area, across Algonquin Park and last year from Fort Albany to Moosonee, down the coast of James Bay. Are those facts right as to your field experience?

A. Yes.

Q. Mr. Macdonald, with respect to the Temagami area in particular, have you travelled and guided in that particular region, extensively?

A. Yes, I have.

Q. Between 1966 and 1970, were you, in fact, based out of a camp in the Lake Temagami area?

A. Yes, I was.

Q. Were you[r] summers spent primarily guiding in the Temagami region in particular?

A. Yes, they were.

Q. Life Project: Are you engaged in an extensive undertaking to record for posterity the native routes of travel and native perception of geography for Northern Ontario and Western Quebec?

A. Yes.

Q. Does this include recording the native names for all the lakes, hills, rivers, streams, rocks and other points of geographic interest to the native peoples?

A. Yes, it does.

Q. In the year 1966, did you begin to concentrate your research to some extent in the Temagami region in particular?

A. Yes, I did.

Q. Have you also researched areas adjacent to the Temagami region?

A. Yes.

Q. In the course of your research, have you worked closely with various native persons and communities to gather information in connection with your project?

A. Yes.

Q. Has your research included communicating with Indians from the: Rama Reserve on Lake Couchiching; Golden Lake Reserve — near Pembroke; Nipissing Reserve — Near Sturgeon Falls; Couchie Settlement — West of North Bay; Kipawa Reserve — on Lake Kipawa, Quebec; Temiskaming Reserve — at Notre Dame du Nord, Quebec; Maniwaki

Reserve — at Maniwaki, Quebec; Town of Mattawa; Ottawa River, including Deux-Rivières and Fort-Coulonge area; Matachewan Reserve near Matachewan, Ontario; Duck Lake Reserve near Chapleau; Mattagami Reserve near Gogama; Biscotasing Settlement on the CPR northwest of Sudbury; Long Lac Reserve near Geraldton, east of Lake Nipigon; Michipicoten Reserve near Wawa on Lake Superior; Osnaburgh Reserve northwest of Lake Nipigon; Fort Hope, Ogoki and Fort Albany Reserves on the Albany River and other reserves in the James Bay Lowlands. And the question that preceded that list was whether or not your research included communicating with the Indians from the reserves on that list?

A. Yes, it did.

Q. Or Indians at those places when the places weren't reserves?

A. Yes.

Q. Does the region contemplated by those communities cover a broad swath of territory totally encircling the Temagami region?

A. Yes, it does.

Q. Have the principal areas of your concentration been in the Temagami area and its closest environs?

A. Yes, that is correct.

Q. In the course of acquiring the data about native routes of travel and the native perceptions of geography, have you, to some extent, gained an understanding of a number of different Indian dialects?

A. Yes, that is correct.

Q. With respect to the native languages, had your ear been attuned primarily in the field, through conversing with various informants, concerning native names for points of geographic interest?

A. Yes.

Q. Have you, in addition, had some formal linguistic training in the native language at the university level?

A. Yes, I have, through Laurentian University.

Q. In the course of communicating with the Indian people you have interviewed, have you encountered spiritual beliefs and ideas that are quite different?

A. Yes.

Q. Have you by and large received such information in confidence and trust?

A. Yes.

Q. In the course of this work, have you interviewed a large number of non-natives, as well as natives?

A. Yes, I have.

Q. Would those non-native interviewees include trappers, government officials, game wardens, prospectors, loggers and, indeed, a host of persons who would have had contact with and information about the natives and the geography of the area?

A. Yes, that is correct.

Q. Was the cross-checking of information from the various sources a significant aspect of this project?

A. Yes, it was.

Q. Would this cross-checking have involved written records, old maps and oral testimony?

A. Yes.

Q. Did you use the cross-checking process to challenge any test data received before accepting at face value what any given informant advised?

A. Yes.

Q. Has your research been carried out totally, completely, and absolutely independent of this land claim?

A. Yes.

Q. Dr. Charles Bishop, earlier in this trial, referred to having examined certain map data.... Mr. Macdonald, do you agree with Dr. Bishop that the map collection he refers to ... contains all of the historic maps pertaining to Canada from the very earliest ones that were produced right through until the early nineteenth century?

A. I definitely do not agree with Dr. Bishop's presumption.

Q. In the course of your project to record the native routes of travel and the native perceptions of geography, have you examined various map collections?

A. Yes.

Q. Would you have examined all of the maps in the same map collection at the public archives of Canada referred to by Dr. Bishop?

A. Yes.

Q. Did you in addition go through other map collections, including all Ontario survey records housed in the Whitney Block under the auspices of the MNR?

A. Yes.

Q. Was that source found to contain historical data of an archival nature not contained in the public archives of Canada collection?

A. Yes.

Q. Did you also go through the map collection held by the public archives of Ontario?

A. Yes, I did.

Q. Did it contain additional historic information?

A. Yes.

Q. Do you have an opinion as to whether or not there are other public and private map collections in Canada not yet culled which may well contain significant historical data to shed light on Northeastern Ontario?

A. Yes, I do. I believe there are other unexplored sources....

Q. Has one primary focus of your research been in completing the interviews with the senior native peoples?

A. Yes.

Q. For what reasons?

A. Since Canada's native people traditionally did not have a written language to record their history, oral traditions are one of the most important sources of historical data. Often whole topics of concern can be successfully researched through oral traditions, where virtually no documentation presently exists. These oral traditions are important because they reflect history, technology, culture, native-white relations, environmental change and concerns from a native perspective. As such, they provide an invaluable addition to the scattered written documents made by Indian agents, priests, fur-traders, land surveyors, timber cruisers, and geologists. Specifically for Northeastern Ontario and Western Quebec within five to ten years most of the senior native people who possess this invaluable information will have died. It is therefore important that the oral traditions be recorded before they are lost forever. This is why my primary focus for research has been in completing the interviews with the senior native people....[1]

Q. Before doing any interviewing, did you previously visit many of these sites by canoe or on foot so that you will have first-hand information?

A. Yes, as much as possible.

Q. Now, I would like you to examine your field interview process. Are most interviews done with the benefit of maps?

A. Yes.

Q. Do you, during the interviews, use different map scales?

A. Yes, I do. My principal scales are 1:50,000, 1:26,000, 1:125,000, 1:506,880. For special areas I have used another five or six scales. I would be dealing with floodline maps, T-plans, et cetera.

Q. Why are different scales required?

A. Work on portages and locations of campsites and trap lines, cabins et cetera, require good shoreline definition. Winter trails require another type of map relying on detailed contour information, so I would use the detailed contour map for winter work. Less detailed maps are used for general orientation and overview.

Q. Approximately, how many working maps would you have, relative to nDaki-menan [the Temagami land claim area]?

A. As I said before, I think it would be working set of about fifty.

Q. Now, in the index [of exhibits] my lord, I had identified Exhibit 19-1 as Nastawgan and Exhibit 19-2 as Geographic names, and the original intent was that the witness would relate one to another to make various points. As it happens, Mr. Macdonald has put it all together on one map, so actually I would rather not change the whole system and I would propose to mark the next Exhibit 19-1 and 19-2 ... Nastawgan and Geographic Names.

Q. What is that exhibit, Mr. Macdonald?

A. This exhibit is a photo reproduction of a map which I used for an overview and for correlating information that I am receiving from various sources and it depicts a swath, it is not coterminous with NDaki-menan, because, of course, I have been working on this a long, long time before even this land claim issue became a current issue in the courts.... [2]

His Lordship: Who prepared this map?

The Witness: I did.

Mr. Clark: Q. How did you prepare it?

A. It is a very complicated and expensive process. First, I did a draft map on mylar, and by photographic process an image of the map was transferred to the gelatin side of a scribe coat. This gelatin surface was then scribed down to three-thousandths of an inch with a scriber and the registered scribe coat was then used to make a ruby lith peel coat. The waterfill areas were then peeled off the ruby lith with micro-forceps. And because of the cost and expense here, what I did was I had this ruby lith peel coat and by a photo-negative process I transcribed that into what you see here. We are looking at hundreds and hundreds of hours of work.[3]

Lifeline I — The Early Years

THE MAN SITTING IN THE SUPREME COURT OF ONTARIO WITNESS box in 1983 was born Craig Kennedy Macdonald on Tuesday, October 8, 1946, in Nettie Koch Maternity Hospital, upstairs at the Alexander Fleury Nursing and Retirement Residence at 5 George Street in Markham, Ontario. Although Markham was still very much a rural community at the time, the City of Toronto, growing north from the old campfires and encampments at the mouths of the Don and Humber Rivers along the shores of Lake Ontario, was pulsing with postwar possibility, reflected in a ten-to-one vote by its citizens in January 1946 to relieve downtown road congestion by building the Yonge Street subway line.

Craig's father, Colin Macdonald, was home after three years of Second World War service in France, Belgium, and Holland with the Royal Canadian Engineers. He was a professional civil engineer with a master's degree in mining engineering from the University of Toronto, and in the tradition of his great-uncle Sandford Fleming was one of the legion of bright minds tasked with figuring out how that subway might run under the infrastructure of Toronto's downtown without shaking the buildings off their moorings. Indeed, throughout Craig's early years, Colin went back and forth from their apartment in the Red House in Unionville to Harvard University, where he was researching the effect of vibrations on built structures with the understanding that this learning could be built into the Toronto subway system from the very beginning to avoid problems and potentially lethal damage in the future.

Craig at about twenty months old cruising on Westacres Drive in the Township of York, where the family lived at number 17 until they built a larger house across the street at number 20.

Learning about Colin Macdonald from Craig's sister, Sandra — rarely from Craig himself — it is clear that the apple had not fallen far from the intellectual tree. "My father came from a long line of geniuses," she said. "Really sort of super-smart people who were in their own little worlds. My father wasn't one of them, but he certainly had a brother who was and there's a cousin who was also very bright and focused."[1] As the book was in process, Craig was a little more forthcoming about his parents.

Craig's mother, Fern Kennedy, was the third of five children of Arthur Coombs Kennedy and Gertie Edna Irene Atkinson. She came into the world on a farm in Edgeley, a few miles west of where she and Colin would begin their life together. But by the time she was ready to go to school, the family farm had been sold, and her father took Gertie and the kids to Hamilton, where he studied to become a chiropractor before moving back to Unionville and setting up a "healing" practice. Fern was fascinated by the alternative medicine, manipulations, and counselling therapies she heard about from her dad. And when she enrolled at the University of Toronto to study psychology and the arts — excelling in extracurricular athletics, particularly the Victoria College women's hockey team — she may have intended to follow in her father's footsteps. But before graduation she fell in love with the best friend of her oldest brother, Keith — Colin Macdonald. It may have been Fern's lifelong interest in the healing arts and alternative medicine that led her to Fleury House, Markham's first pre-hospital birthing centre, instead of Toronto General Hospital or a more formal medical context, to give birth to her first child that October day in 1946.

College sweethearts, Fern and Colin were in their midtwenties when they married in 1938. Where Fern, who was three years younger than Colin, had been more of a homebody, content to take clerical jobs close to home when she finished university, Colin was a bit of a rover. When he graduated from U of T, he took a job at the McMillan Gold Mine, at House Lake southwest of Espanola. Following this he had mining engineer positions at Falconbridge Mine in Sudbury, E.B. Little Prospecting in Haileybury, Matachewan Consolidated Mine in Matachewan, Sylvanite Mine in Kirkland Lake, Paymaster Mine in South Porcupine, and, finally, at the Department of Highways in Stratford. Colin procured a prospecting licence as well and did a little mineral exploration in his spare time. Interestingly, Colin was also involved in the dewatering of the Golden Rose Mine in Temagami a few years prior to Andrew and Larry's fateful *War of the Worlds* trapping trip.

When Fern was with Colin in this adventuresome life, both of them thrived in northern Ontario: canoe trips down the Mattagami River and lots of cross-country skiing. Fern was also pitcher for the women's baseball team that won the provincial championship. And even though prospecting was not her thing, the two of them did some rockhounding from time to time.

But even before that, Fern's older brother, Keith, also a U of T mining engineer, teamed up with Colin on university breaks. One year, they drove Colin's twenty-seven-dollar used Ford Model T across North America. They shovelled coal or fed cattle for free passage across the Atlantic to explore Europe. And, on another occasion, they were hired by a lady in Toronto to drive her, in her luxury Pierce-Arrow roadster, to California, returning on a grand adventure hitchhiking and hopping freight trains up through British Columbia and back east, which took them to the 1934 Chicago World's Fair before returning to their studies in Toronto.

Any notions of Fern and Colin continuing these adventures by having a family were swept aside by the Second World War. Even though Colin was quite a bit older than the usual enlistees, his skills as an engineer were courted by the Royal Canadian Engineers, resulting in an officer's commission in 1942 and a free pass to France and northwest Europe for the remainder of the war. With her newly minted B.A. from U of T, Fern took a secretarial job with the Canadian Broadcasting Corporation. Communication with her new husband narrowed to the occasional letter from overseas. Still living in rented accommodations in Unionville, Fern was able to enjoy the company of her family while living out the war years on her own.

On his return from the war, the adventure itch being well scratched by his service, Colin settled into nine-to-five employment, taking a city job where he joined those working on the original plan for the Toronto subway. Craig's arrival in 1946 brought with it a need for additional space. And so, as their firstborn was finding his feet, the Macdonalds moved to a modest bungalow at 17 Westacres Drive in the Township of York where their second son, Roderick, arrived in 1948, followed by their daughter, Sandra, in 1951. Again in need of additional space to accommodate their expanded family, Colin and Fern were inspired by the spate of building that was turning farmers' fields into neighbourhoods around them. Eyeing a For Sale sign across the road, they eventually bought the lot and built a house at 20 Westacres Drive, into which the five of them moved in 1952.

The children's school, Charles E. Webster Public, was just around the corner, close enough to walk and slip home for lunch. It was a neighbourhood full of young families and semi-feral kids ready for baseball, road hockey, hide-and-seek, kick the can, whatever. A couple of blocks to the west was Black Creek and the greenbelt that wound south to the lake through Keelesdale Park. Nearby was a host of other safe places to run and explore.

Part of the magic of the family cottage on Lake Kashagawigamog in Haliburton County was the chance to "mess around in boats." Here is Craig and his younger brother, Rod, trying out their helmsmanship in the 1950s.

On weekends, it was visits and meals with Art and Gertie, their maternal grandparents, who lived in Unionville. And on school holidays, they would load up the Nash Rambler and head to a simple family cottage on Lake Kashagawigamog in Haliburton or to Leamington to visit their grandmother, Annie Goodchild Macdonald, making side trips to visit Great-Uncle George Macdonald in Windsor, but that's a whole separate story. And so unfolded an idyllic *Swallows and Amazons* situation, very similar to the magical world of outdoor adventure created by British children's author Arthur Ransome, for Craig and his siblings.

By all accounts, school was something that Craig tolerated, doing what he was told, but it was never a great fit. From Charles E. Webster Public School to York Memorial Collegiate Institute, it was all a bit of a slog. "Everything that he was good at," said Sandra, "was not taught at school."[2] What Craig was good at was marching to his own drum with extracurricular dreams and projects of his own. Kites became a big thing for him. Small schoolyard kites were quickly replaced by larger homebuilt kites of many shapes and designs. First came small cruciform Eddy kites he flew in the schoolyard. Then, aided and abetted by friends and his dad, who helped with getting the materials they needed, small kites became bigger *W* kites and eventually box kites — serious box kites, the largest of which was eight feet long with a twenty-foot tail and flew on ten thousand feet of line!

To achieve great altitude, Craig and his pal Juris Veveris, with the assistance of other neighbourhood kids, flew a train of three large kites on the line, each one reaching higher than the one behind. At one point, Craig got the idea that some aerial photography might be a good idea, so in another plan created and executed from his bedroom at 20 Westacres, he fitted one of those big box kites with a Brownie Starflash camera using an aneroid barometer to control shutter release. History doesn't record whether the black-and-white snapshots, when Craig and Juris picked them up at the Rexall Drugstore on Eglinton Avenue, were any good. Craig concluded that hydrogen-filled balloons would be better for air photography and launched a few of those as well.

They flew kites in the light and in the dark, in one instance hoisting a railroad flare into the night sky so high that people called the police with UFO sightings. And in another instance of Macdonald engineering, Craig solved the problem of knowing what kind of wind was blowing at altitude, especially during night. He created a mechanism by taping narrow strips of kitchen tinfoil onto a small rotating anemometer attached to a battery and a little flashlight bulb. Every time the propeller on the anemometer rotated, it caused the light to flash, which allowed kite pilot Craig to know, depending on the flash rate, if there was still enough wind to keep the kite aloft.

Crashing kites were always a serious concern, because these creations were works of art and ingenuity that took hours to make, but especially so after one of his smaller kites, on a long line, dived unexpectedly and clobbered an unsuspecting police officer leaving the nearby precinct on foot. That they didn't get into more trouble with all of this aerial activity so close to Downsview Airport, barely three miles north, is a bit of a miracle. It was probably just as well that Craig's experiments in public school with making two small gliders were never flight tested. The first was destroyed in a windstorm and the second one's wings were never covered with fabric. However, during high school he did manage to build a lighter, more flight-worthy hang-glider — long before the hang-glider craze got going in California. With four people pulling on a tow rope while riding downhill, Craig did manage to take his brother Rod and another neighbourhood kid on several flights. These were short hops of about a hundred yards or so off the bluffs at Black Creek.

School notwithstanding, the late 1950s were heady times for anyone with scientific curiosities, like Craig, because this was when the space race was beginning. In October and November of 1957, as Craig was turning eleven, the Russians flew Sputnik 1 and 2, the earth's first artificial satellites. The papers of the day were full of comments about how the Russians had beaten the Americans into space. Whatever Craig might have been reading for school — or whatever he might have been *supposed* to be reading — he was certainly well aware of the Sputnik and Vanguard rocket programs, and at some point around this time, he took it upon himself to build his own rockets. Craig clipped newspaper and magazine articles about rockets and pasted them into a scrapbook, which he still has to this day.

School, the public library, advice from Juris's mother, who was a chemist — these were all sources available to Craig as he and Juris went about making home-brewed gunpowder to power the rockets they would build. For carbon, they took lumps of readily available coal — anthracite — that people were still using to heat their homes in those days. With a file and some elbow grease they created powdered carbon. To that they added powdered sulphur, readily available at the local gardening shop, and saltpetre (potassium nitrate) that likely came from Rexall Drugstore. They mixed all these together and stuffed the fuel into rolled paper tubes they fashioned with nose cones, fins, and nozzles — just like the real rockets. They made fuses by rolling a little of the gunpowder in toilet paper, and headed off to the schoolyard to try them out.

The early rockets worked, but Craig being Craig, he wondered how they might make the engines more powerful. So they started improvising with the ratios of the three reagents in the fuel until they got the idea that changing from saltpetre to the more reactive potassium chlorate and manganese dioxide. Where they got manganese dioxide and potassium chlorate is anyone's guess. Unlike saltpetre, which is used with table salt for curing meat, these two chemicals are not commonly stocked in local stores. When they started mixing up this new batch of fuel, it became hot to touch, and they even started to smell chlorine gas, so they put the fuel in Mason jars, which they immersed in a laundry tub of cold water to keep it cool.

At the same time, Craig decided that paper tubes weren't strong enough to realize his rocket's full potential, so they got a lightweight aluminium tube — a short piece of pipe — and fixed this up with a launching stick, fins, nozzle, and an upper plug. While loading the rocket, it was buried in snow to keep the fuel as cool as possible. At this point, Juris and Craig had misgivings about proceeding further, but they were outvoted by the rest of the group of neighbourhood kids who had gathered to spectate.

On what became a fateful day, and the sudden end of Craig's rocket science career, with the new super fuel loaded into the metal rocket, they headed to Webster schoolyard for the launch. This time, though, instead of elemental carbon they made themselves, they'd added a new variable to the mix: a box of soot from the school's furnace that they'd found in the garbage.

On this occasion, Juris and Craig were joined by a new kid called Roy Culpepper, whose family had recently moved to Toronto from Pakistan. Roy volunteered to carry the rocket over to the schoolyard. It was feeling hot as they made their way around the corner, so they put more snow around it to keep it cool, pre-launch.

Roy had just slid the rocket under the school's chain-link fence and was getting ready to climb over himself when — *BOOOOOOM!* — the rocket exploded. The force blew a hole in the chain-link fence. A piece of shrapnel from the fiery hot metal tube took off half of Roy's index finger and carried on through his arm and into his leg. Roy was totally stunned but still conscious, bleeding from his hand, arm, leg, and his nose. His glasses were totally shattered. His pants were in shreds. And one arm of his winter coat was in ribbons. Craig was twenty feet back and wasn't actually looking at Roy when the explosion happened, but when the dust settled and Roy had been taken to hospital by ambulance, Craig realized that another piece of shrapnel had cut his lip and chipped one of his own front teeth.

Roy survived and went on, with his nine remaining fingers, to complete a doctorate in economics at the University of Toronto. Craig, on the other hand, answered all the police officers' questions when they came to the house later that day to interrogate him about the accident that had nearly killed the three of them. He told them that he thought it might have been impurities in the soot from the furnace that compromised the fuel and ultimately may have caused its spontaneous and premature ignition. Having unwittingly created a pipe bomb, this seemed like a good moment to turn to other, safer, projects, like building soapbox racers or flying his younger brother off the Black Creek bluffs on a homemade hang-glider. Craig does remember that the police, as they scolded him, were somewhat incredulous that a bunch of Grade 8s could know this much chemistry, and about the rocket reaction, and especially about the chemicals and where to get them.

What Craig did not attain in his school studies by way of confidence, drive, and application he certainly made up for out of school. There were family road trips around the Gaspé Peninsula to the Maritimes and, in another year, to Manitoba. There were also two-brothers-prospecting trips with their dad, one to Killarney and the north shore of Lake Superior, which took them to what was then the end of the Trans-Canada Highway at the Agawa River north of the Sault. On the other side of the river, road crews were active, and Craig remembers that their dad, the civil engineer, turned to him and Rod and said, "Boys, it's going to cost a million dollars to get up that hill."

There's a particular memory of summer prospecting on the north shore of Lake Huron, stretching their long Macdonald legs on outcrops in Rainbow Country looking for arsenopyrites, which are often indicators of nearby gold or silver deposits. There's no doubt that both Craig and Rod picked up terms and processes from watching and listening to their dad as he did his actual prospecting, but Craig remembers another aspect of this journey, which was stopping in at the Chief's Office at Whitefish River First Nation to ask for permission to look around on the rocks at Birch Island. This would have been a much more normalized encounter — no feathers or buckskin outfits — than Craig might have imagined from the stories about Ernest Thompson Seton's Woodcraft Indians that he might have read or heard about in Cubs.

Craig, no doubt, would have asked about the incredible clarity of the lakes in the Killarney area, which were set into the spectacular quartzite rock of the La Cloche Mountains. His dad may well have talked, even in those days in the late '50s or early '60s, of oxides of nitrogen and sulphur from upwind smelters combining with water in the air to form nitrous and sulphuric acid. Without any buffering from the quartz rocks, which is what you would get in limestone country, this would increase the acidity of the lakes to the point that almost nothing organic would grow in them. Almost since the womb, Craig was fascinated by how things work, from kites to chemical reactions.

These times in the car, whether with the whole family or on road trips with just the Macdonald boys, always gave their dad the chance to tell stories of his exploits growing up and with their uncle Keith and, of course, of his experiences as a member of the Royal Canadian Engineers in the Second World War. Tales were also told while camping, at the cottage in Haliburton, and at the kitchen table. Rod and Craig — both of whom had capacious memories for the details — thrived on these stories.

There was a basement rec room at 20 Westacres, a place where the three kids could let off steam or while away rainy Saturdays. It was lined with Colin's reconnaissance maps from the war. There were board games, boxing gloves, a dartboard, a piano (Rod was accomplished), and other toys that spoke of growing minds and strengthening bodies, but it is easy to imagine Craig getting caught up in his own little reverie from time to time, pondering his dad's army maps and tracing lines that marked some of the bombed roads and bridges Colin was tasked to reconstruct after D-Day.

The only home that was more interesting to Craig was his great-uncle George's place in Windsor. Son of a mercantile family who owned a big department store — Bartlet, Macdonald and Gow — in downtown Windsor, their grandfather William Alexander Macdonald had been a medical doctor and travelled the world, learning other languages and getting extra training in his ear, nose, and throat specialty in places like Germany, Scotland, and France. Through his twenties and into his thirties — 1895–1910 — William Alexander Macdonald's passions and curiosities kept him on the move. When he did marry Craig's grandmother Annie Elizabeth Goodchild, their house in Windsor was full of mementos and stories drawn from a lifetime of travel and adventure.

George followed the more conventional vocational path and took over the running of the store from his father, but his avocation, his absolute passion, was for history, archaeology, and anthropology. His basement was festooned with treasures he had unearthed during digs and explorations he undertook outside of his duties at the store. There were flints, arrowheads, muskets, cannonballs stacked up, axes, maps, charts, letters, "tons of stuff," Craig told me, all displayed with notes and details that revealed Great-Uncle George's encyclopedic knowledge of antiquity. George died in 1959, and his collections became significant acquisitions for the Hiram Walker Historical Museum (now in the Archives of Ontario), the Windsor Municipal Archives, the François Baby House museum in Windsor, and also Fort Malden National Historic Site in Amherstburg.

For Craig, and to a lesser extent for his sister and brother, Great-Uncle George's basement was a magical place. His notes, field sketches, maps, and stories captivated Craig through his early years and into his teens, perhaps the most formative years of his life. Together they would travel back in time and onto the landscapes of Ontario and beyond. The artifacts were physical touchstones for Craig to ways of life that connected their makers to the resources of the land. He would look and listen and always had questions, lots of questions, to which Great-Uncle George would provide answers from his fertile mind, as if that was the only thing he had on for the moment, for the day. "I loved talking to him. He was a very kind and generous man. He was a huge influence on me," said Craig in later years.

It may have been Great-Uncle George's inspiration that kindled Craig's interest in sketching and art. With lacklustre engagement in school, anything that Craig showed a particular aptitude for was supported by his parents. In Grade 6 or 7, his mother found an art teacher, a Dutchman called Tye Hornfeldt, who for a year or so engaged Craig with oil painting lessons that became something of a foundation in Craig's later life. Plans for kites, soapbox races, and rockets were all enhanced by Craig's eye-to-hand facility with drawing, particularly sketches that showed what connected to what and how things worked.

These were usually done on foolscap in pencil. Difficult to say why this combination of lead pencil and distinctively sized lined paper was Craig's go-to for his notes and sketches, but this is *all* he would use for observations and interviews for the remainder of his life. Perhaps the highlight of his art career was a cartoon he drew when the Beatles came to Toronto in 1964 that was published in the *Toronto Star*. It was a scene with a whole bunch of heads and a sign that said "Beatles Fan Club." Everyone in the image had a Beatle haircut except for one bald guy in the middle, who had two hairs sticking out of his pate.

Craig moved from Webster Public School to York Memorial Collegiate Institute, a longer walk west. Trying to sit still in class and learn French, Latin, and German was a trial, but middle- and long-distance running helped make school a bit more appealing for a while. He was a member of the York Memorial team that set a regional record for the 440-yard relay race, and in the all-Ontario competition in 1960, Craig came fourteenth in cross-country, running in a field of two hundred competitors. But Craig made it his business to figure out who was winning and why, and at some point around Grade 11 he decided that he didn't have the lung capacity that some of his competitors did — that was just a fact of his life — and there ended his running career.

There were two teachers to whom Craig was drawn, Mr. Irwin at Webster and a chemistry teacher at York, because they both had special affinities for Indigenous Peoples and the northern landscapes of Canada. And a Grade 9 math teacher who embroidered his algebra lessons with stories of working as an itinerant teacher on the railway, teaching kids up and down the line in northern Ontario. But beyond that, the most transformative part of Craig's education happened outside of school — through his family for sure, through his own neighbourhood ramblings and projects, through various summer camp experiences, but also, significantly, at the Macdonald family cottage in Haliburton.

At the cottage on Kashagawigamog Lake, Craig could move his technological improvisations from home into a semi-wilderness setting. What was different at the lake was that his parents — and often his maternal grandparents, Art and Gertie, and other relatives — were there on holiday. They had time to do things with the children, as well as projects of their own through which they could teach the children, particularly Craig, who had a natural affinity for handwork and storytelling.

One such project, undertaken with his grandfather and father, involved using dynamite to build an all-season road into their cottage through the unforgiving granite of the Canadian Shield. This experience led Craig to a lifelong interest in trail building and to his eventual retirement from the Ontario Ministry of Natural Resources (OMNR) as the "last licensed blaster in Algonquin Park." Six-year-old Craig would hold a metal drill steel to make a hole in the rock for sticks of dynamite while his father, uncle, and grandfather took turns pounding on the drill with ten-pound sledgehammers. After each round of hits, Craig would turn the drill steel a quarter-turn and blow the dust out of the deepening hole with a drinking straw. As the hole deepened, Craig had to rotate a special spoon device mounted on the end of a metal rod to keep the hole clean of dust.

When the actual blasting was being done, there was a protocol to follow because the owner of the land, Gord McManus, had "shell shock" (perhaps post-traumatic stress disorder in today's parlance) from artillery service in the Second World War. They would notify him so that he might head to the Dominion Hotel in Minden for a few pints to avoid hearing the percussive blasts from making the cottage road.

Of many formative and foundational experiences at the cottage, cutting wood with his grandmother was among the best and longest-lasting lessons. "She could manipulate a crosscut saw," he said with great affection. Because she'd grown up on a farm in Edgeley, it might be easy to imagine Gertie, who Craig called "Grannie," milking cows and splitting wood for the kitchen stove. "But," said Craig, "she was so good at using a crosscut saw. So good. She knew how to lay one hand on top of the blade and push with the other just enough to help me without kinking the blade. Like real nineteenth-century pioneers stuff." Craig took these lessons with him into his various jobs with the OMNR, including his work as part of the field instructional staff for the OMNR Junior Ranger program. There he taught young city kids, just as he had been, the difference between teeth that did the actual cutting (which had to be angled just so to do the maximum amount of work with the minimum amount of effort) and the rakers that cleared the kerf (as if the junior rangers would have any idea what that was) of sawdust so as not to bind the blade.

One thing cottage life did not do was breed independence in a young man's life. So as early as 1953, when Craig was seven years old, Colin and Fern sent him to Camp Wawaganoma, a York Township day camp in Keelesdale Park. Overnights in wall tents were first on the Humber River but later moved to Sixteen Mile Creek. There he learned basic camping and campcraft skills — making knots, fire-lighting, woodcraft, axe care, tent-pitching, cooking on a campfire, canoeing, swimming in the river, and camping overnight in canvas wall tents, and on day trips he practised canoeing in tippy Sunnyside Cruisers behind breakwalls on Lake Ontario.

Sandra thinks Craig's first paddling experience might have been at the age of four or five, sitting on a log with his grandfather, transporting it down the lake to build a dock, but Craig is convinced that he got his first taste of canoeing with his counsellor and fellow campers from Wawaganoma, when they went down to the beach at Humber Bay on Lake

Both Craig and brother Rod took immediately to life in a canvas wall tent at Camp Wawaganoma on Sixteen Mile Creek in York Township.

Ontario and played in Sunnyside Cruisers for the afternoon. Whether at camp or at the cottage, canoes entered Craig's life early in his formative years and remained a motif — as a vessel, as a way of thinking about the world, particularly about the relationship between people and the places they live — for the rest of his life.

Three years of summer day camp led to more independent overnight summer experiences at Camp Sharbot, near Sharbot Lake, in his middle-school summers, including his first canoe trips in the Rideau Lakes. Rod and Craig, with their belongings in canvas duffels, boarded the steam train at Leaside Station and headed three hours east. A big adventure at the time, to be sure, but one that mapped quite nicely onto stories of their family members doing exactly the same decades earlier. Little did Craig know that the routes, the lakes, and the campsites imprinted in his almost photographic memory would become part of the mental maps that would guide him as he spent a substantial part of his career working for the OMNR in that very park.

Camp Sharbot gave Craig what he was hungering for: campcraft; tenting; canoeing and canoe tripping instruction; overnights in tents on the Fall River, travelling in canvas-covered wooden canoes; and chances to try out these skills in new and challenging circumstances. That drive to face a challenge cut short his time at Camp Sharbot when, as he was starting high school, Craig and Rod enrolled at Camp Kandalore in Haliburton. Although, in different ways, these summers transformed their lives.

• • •

In 2020, as I developed the research for this book, Craig recounted his move from Camp Sharbot to Camp Kandalore. The details of his memories — sixty years later! — was granular, indicating the power of the experience. In 1959, Craig was twelve, turning thirteen in October. Even before he got to Camp Kandalore, he told me, the Haliburton landscape was present in his mind's eye because of stories his dad had recounted. Colin had worked as a young engineer at a survey camp on the Gull River, just a few miles from Kandalore.

Colin had taken the Victoria Railway north from the main CPR line at Lindsay to Gelert Station, and then travelled on a buckboard wagon to Minden and beyond to the University of Toronto Survey Camp. He told his son about the camp life, dances at Maple Lake, and the legendarily bad roads through the hard rock of the Canadian Shield in Haliburton County. Craig's memory, uniquely tuned for geographic detail, had logged all the places and names in his father's stories, which allowed him to situate Camp Kandalore in the growing cartographic database in his young head.

"My first year at Kandalore," Craig said, "I was Intermediate 2. My counsellor was Bob Govan. He became a medical doctor, practised up in Timmins and Marathon. My section director was John Swinden, who was there for a couple of years, 1959 to '60. Nice guy who came from Fort William. His father worked for Nipigon Corporation in the accountancy end of things. Both Bob and John had excellent bush skills. I was as good with my bush skills as anybody else in my tent."

Craig, although he would be loath to admit this, harbours a pretty serious competitive instinct. Perhaps he arrived at Kandalore that first year feeling like a bit of a tourist, coming from a less woodsy camp. It must have been a relief to him that his bush skills, learned at day camp and improved at Camp Sharbot, were as well developed as those of any of his tentmates who had grown up in the Kandalore ways of doing things. But it's clear that he thrived at Kandalore from the moment he arrived.

"My first canoe trip at Kandalore was in the first week of July that summer," Craig recounted. "We went from the canoe dock on Lake Kabakwa, where the camp was situated, through Lake Kushog, then through Raven, Fox, Kawagama, Slipper, Stocking, Havelock, Johnson, Kelly, Kennisis, Red Pine, Clear, Big Hawk, Sherborne, and back. I can tell you where every campsite was. Our first night was at Tyrell Island on Kawagama. Second night was on the outlet between Kelly and Johnson, on the right-hand side. Third night was at Clear Lake. The fourth night was on Sherborne, the east end on the right-hand side, close to the dam. That's the dam that goes into St. Nora."[3]

Damn! What a memory.

Craig did so well at Kandalore that his parents enrolled him for one month in the succeeding summers. He worked his way up from the Intermediate to the Senior Section of the camp, each increment involving longer and more remote canoe trips that moved from the Haliburton environs of the camp itself into Algonquin Park and eventually into the wilds of Temagami. To this day, he carries with him instant recall of the encyclopedic detail of where they went on each trip, who was on the trip, whom they met along the way, what the weather was like, and whose sleeping bag caught on fire by accident.

And it wasn't just canoe trips. At the time, Kandalore was experimenting with hiking expeditions as well. In Senior 1, Craig remembers getting dropped off at Raven Lake with his counsellors, Murray Morris and Tiny Taylor, and paralleling on foot the canoe trip he had done as an intermediate camper, bushwhacking all the way down the back side of Plastic Lake to Sherborne Lake, across the dam, then to Silver Buck, Silver Doe, and along the cliffs of Kushog Lake and back up the road to Kabakwa Lake and camp. By the time Craig was seventeen and old enough to join the camp staff as a counsellor in training, he had the maps of Haliburton and Algonquin Park and the lands in between indelibly etched in his memory.

And unlike his continuing undistinguished academic and athletic career at York Memorial Collegiate Institute, Craig was building at Kandalore something of a reputation for himself as a resourceful and tough-as-nails tripper and woodsman. Legend had it that he once started a fire from scratch with a friction bow drill in thirteen seconds, drawing on training from his father, who had superb woodsmanship skills. Likewise, when it came to axemanship, he lent his treasured modified Walters Axe Company personal axe to a camp colleague, who beat Kandalore's brawny owner and director, Kirk Wipper, in a chopping contest. After defeat, Kirk tried Craig's axe and then, in an uncharacteristic fit of temper, claimed that he had been cheated.

To conserve a dwindling supply of matches, Craig taught himself to start four fires with a single Eddy match. With his razor-sharp belt knife, he removed the match head and then quartered it. The cuts were made very slowly to prevent premature ignition of the red and blue ends containing sulphur and potassium chlorate, which he knew about from his rocketry days. Each quarter was then placed on its own pile of tinder. He would make a garrotte with three feet of stove wire and sticks, which he would then wire until it was red-hot by rubbing it back and forth on a piece of dry wood. He then used the hot wire to ignite the quarter matches, one at a time, into four separate fires. Magic. Even if the match-head chips were soaking wet, they could be steamed dry with the hot wire before ignition. Little wonder that somebody — it might have been veteran tripper Mickey Houston — started calling him "Sticks" or "Styx" Macdonald, a nickname that follows him to this day in Kandalore circles.

One of the bizarre traditions (now for sure, but also at the time) at Kandalore and many other children's camps throughout Canada in the postwar years was "Indian Night," an homage to the original inhabitants of the lands in which they made their canoe trips. These invariably involved directors, like Kirk Wipper at Kandalore, dressing up as the "Chief" in full buckskin regalia with showy Plains headdress, complete with beads and feathers, and convening campfire gatherings at which stories would be told and traditional games played that evoked some western cigar-store version of what it meant to be an Indigenous person in Canada. Campers would be encouraged to "Indianize" their names. In their cabin groups, they would apply craft "war paint" to their faces. They'd grab a blanket, and with bare chests and breechclouts fashioned from old swimming towels and a piece of knotting cord or a belt, head to the Indian Council Ring.

Craig, whose ancestors had a nearly three-hundred-year association with Indigenous people, was never able to find any enthusiasm for Indian Night. Instead, perhaps predictably and probably totally unwittingly, Craig burnished his growing stature as a campcraft virtuoso by organizing the alchemy behind the fire that miraculously burst to life at the "Chief's" will and command. Already well versed in combustive chemistry (with a chipped tooth to prove it) and working with the Kandalore program director of the day, John Denys, Craig buried a small-diameter pipe or piece of conduit leading from outside the council ring to the centre of the hearth in the middle of the ring. Through this pipe he threaded a tripwire. Then, hidden under the pile of firewood, he placed a dish of icing sugar mixed with potassium permanganate. On top of this, he placed an open bottle of nitric acid that could be tipped over with the tripwire to ignite the mixture, causing instant spontaneous combustion to ignite the firewood.

The "Chief," in his full headdress, would arrive at the Indian Council Ring at dusk in his favourite Indian Maiden Chestnut canoe, often standing in the front and being paddled by a trusted "brave." He would solemnly step out of the canoe in his moccasins, silently survey the assembled company, and then step forward with opening remarks, which may well have included words spoken in the Haida or Ojibway language. Raising his arms, calling in the spirits of the four directions, particularly the Thunderbirds from the west, he would call for the council fire to commence. Craig, who was crouched in the bushes just outside the balsam-pole palisade walls of the Indian Council Ring, would pull the wire, and *poooof!* the fire would magically spring to life.

However well-intentioned this type of charade, it continued to misrepresent and diminish Canada's Indigenous Peoples, right up into the 1980s. At camps like Kandalore, it often included special guests. Kirk Wipper seemed to have a broad network of contacts who would come to the camp to help with programming. Craig remembers one special guest, who was blind, reciting a canon of Robert Service poems at Indian Night. Another visitor was a local trapper, Hugh Taylor, who spoke a little about trapping. Another guest was a champion log-roller. And on one very special occasion in the early 1960s, the guest at Kandalore Indian Night was Bernice Loft Winslow, also known as Dawendine, daughter of Six Nations Chief William D. Loft. In the tradition of Pauline Johnson, she read from a book of her poetry and spoke of her desire to help camp directors like Kirk to "indianize" Indian Night with the stories and sensibilities of authentic First Nations Peoples. Craig was intrigued and yet was never much for ceremony of any kind, so regardless of what others might have thought about Kirk's special Mohawk guest, Craig stuck to his helping role in making sure the fire got started on cue, without a hitch.

Perhaps most indicative of Craig's distinctive approach to life and living was that in free moments at camp, when others were napping or reading comics, he loved to climb to a lookout at the end of a trail he'd found as an intermediate camper. The trail started behind the staff cabin called "Eagle's Nest" in the Intermediate Section and led up through a forest of mature red and white pines to a high point behind the camp, from which he could see for miles out across Kushog, the neighbouring lake.

In his counsellor-in-training (CIT) year, he took a telescope that Great-Uncle George had given him up the trail. Through the lenses that had brought him his first glimpses of the moons of Jupiter and the Andromeda Galaxy, thanks to the backyard teachings of Great-Uncle George, Craig reacquainted himself with a great white pine on the far western horizon that stood "head, tummy, and shoulders" above the surrounding forest. As a kid, he'd promised himself that one day he'd walk to that tree. And during his CIT year, he made good on that promise. He recounted:

> On a day off, I made the trip. I didn't know how far it would be. I took a can of beans, an axe, and a blanket and took off into the bush. I would go over a few hills and then I'd climb a tree and see the tree off in the distance and I would just keep going farther and farther. I'd go another few ridges and climb another tree and I'd still see the tree. I finally got to the tree and got out my axe [to make a little blaze]. And wouldn't you know it, I left my little axe at the base of that tree. It was a crummy old thing. It was half ground down.

So there, at the base of the biggest old white pine in a remote corner of Hinden Township, on the western edge of what is now the amalgamated township of Minden Hills, was Craig Macdonald's much loved Walters hand axe. Maybe it's still there, as a testament to the tenacity of a guy who would get a bee in his bonnet to do something and then stick with it, through thick and thin, until the goal was achieved, until the job was done. Getting to that tree, however, had taken up most of the day. The sun was starting to set and Craig had to get back to camp so that he would be back on the job the following morning.

Recalling this crazy Don Quixote venture, he described what happened on the return journey:

On the way back, I had to go through that huge bog that's back there and I was losing light. Eventually, when it got dark, I had to stop, so I lay down in my blanket at the edge of a beaver pond. Through the night the beavers would come up to me and splash, make a big splash, and wake me up. It got cold. It was a clear night. My one blanket was just not adequate. The beaver pond water wasn't very good to drink, so I was kind of parched, having walked all day. I remember getting up and walking through the fields of that pioneer farm that's back there, walking through fields watching the sun come up. But I made it in for breakfast. I was pretty hungry, but I got to see that magnificent tree.

Travelling with Craig I — Kandalore Days

FULL DISCLOSURE, I WAS ONE OF THE SEVEN-YEAR-OLDS WEARING war paint and a breechclout, sitting with my counsellor and junior cabin group in the Camp Kandalore Indian Council Ring on a night in 1963, amazed by the way the "Chief" magically summoned the fire. We had all practised the toe-heel, toe-heel Indigenous dance step that we had been taught by our campcraft instructors and, in our moccasins or bare feet, we had paraded through the main door of the palisade and circled the fire, which was set but not burning, before taking our seats to await the commencement of the festivities.

Although a parody of the highest order (looking back I cringe to think that this was *ever* even going on) the whole affair had a kind of solemnity to it. The message, much like the feel of crossing the rickety wooden bridge to Chapel Island on Sunday mornings in the whitest T-shirts and shorts we could muster, was that this remembrance of the first inhabitants of the woods we were learning to love was a semi-sacred observance. But it was a little weird to be dressed the way we were, feeling the cool air of the evening in places normally covered by garments like shirts and underwear. If I did see Craig Macdonald hunched in the bushes outside the council ring, awaiting the signal to pull the tripwire on the chemical fire starter, he was very much upstaged by the mysteries of what we were to behold inside the palisade walls.

Although I had not yet met Craig Macdonald in my time at Camp Kandalore, I knew who he was. Because, even as a CIT, he carried with him woodsmanship distinctions from his camper days that set him apart from regular campgoers, who did not have the patience or the determination to excel at tasks like axemanship and matchless fire lighting. When he was a senior camper, Craig was the guy to beat in competitions in just about every category, on land and on water.

Craig had mastered log-rolling. At the waterfront, between the canoe dock and the main swimming area, were a couple of stout white pine logs the length of a pickup truck and perhaps a couple of feet in diameter. These had been peeled and longitudinally grooved with a scoop chisel (maybe by Craig himself — he was keen on such details). When not in use, these logs would be cranked out of the water by an old logging winch, but on regatta days and for special woodsmanship competitions, they would be floated so that all comers could try their skill at log-rolling. With his bandy white legs, arms, and torso (because he was forever in work pants and long-sleeved shirts and never one for gratuitous tanning), and wearing a style-be-damned bathing suit, Craig would prevail. It was as if he had an inside track to the focus, concentration, and balance it took to win at log-rolling.

My first memory of actually meeting Craig was on an overnight on Kabakwa Lake. The camp's raison d'être was canoe tripping, so everything presented in the daily programming schedule was in service to building the skills and knowledge necessary to venture out by canoe on multi-day trips, first in Haliburton, then moving to Algonquin Park, then north to Temagami, and beyond to the big rivers flowing to James Bay. But we had to start somewhere, and that was on overnight campouts right on the camp's home lake.

As a CIT, Craig was, I suppose, assigned to our counsellor or to the Junior Section of the camp, and so it was that he joined a couple of cabin groups that boarded two big Couchiching Freighter canoes for a paddle to a campsite on the other side of the lake. There we would set up our tents, unroll the waterproof bedrolls we'd learned to pack in our in-camp lessons, and then go through the routines of firewood collection, camp cooking, cleaning up, safety, environmental ethics, all of that. It was also part of the camp tradition to tell stories around the campfire after dark — and that's the memory of Craig Macdonald that lingers.

It was a story about a young couple who were buried in an avalanche in a brand new Ford Thunderbird. It was a gleaming cream-coloured hardtop, with white leather seats, serious whitewall tires, and a 390-cubic-inch V8 motor with a three-speed Cruise-O-Matic transmission. From the moment he started telling the story, it was the details that commanded our attention. This guy knew his cars, and he knew his mountain topography.

The couple were travelling on a deep mountain pass somewhere in B.C. The snowpack had been unstable that year. Avalanches were always a thing in the Rockies, he told us, but some years all it would take was something like the reverberation of a big V8 motor to trigger a slide. And that's exactly what happened. Lovers buried alive in a steel, glass, and leather tomb, courtesy of Henry Ford.

The story, as one might imagine, boggled our seven-year-old imaginations, particularly the part about the brand new car being dug out of the ravine. But instead of dwelling on the unfortunate couple who died, the tale followed their ghosts, who stayed with the car as it was cleaned up and offered for sale at a used car dealership in Penticton.

Sadly, the details of Craig's story from there are not clear, lo these sixty years later, except to say that he recounted how the car was still a stunning sight, except for some staining on the leather that the detailers could not conquer. And, strangely, every prospective buyer who sat in the car or took it for a test drive was mysteriously compelled to say no to the fancy cream T-Bird and keep looking. Craig thought it might be the smell they could never get out of the leather, or maybe something else. And I'm not 100 percent sure, but the dealer, loving the iconic car, might have driven it himself and died in a horrific car accident.

Something about Styx Macdonald — and the way he told the story — pointed to the undeniable fact that he was a bit different than the other camp counsellors and instructors. He didn't laugh very much. Not because he didn't have a sense of humour — he did. He laughed at things others didn't find quite as funny. It was that the things he was interested in and the things he had to say seemed very important. Whether it was getting the minor details of the story right (even though I'm sure, looking back, that he was drawing most of them from the depths of his imagination), relating the history of how campcraft skills came to be, or going the extra mile with his teaching.

With axe instruction, for example, it wasn't enough that Craig show campers how to swing an axe without embedding it in their own foot or someone else's. Safety was important. He surprised me — and, I presume, others — with his tutelage that a dull axe is more dangerous than one sharp enough to cut paper or shave with (although most of his students in the Junior Section knew little about shaving). And he needed to tell us that when it comes to putting that edge on the axe with a file, a whetstone, and a leather strop, the angle of the cutting edge is sharper for a cutting axe than it is for a splitting axe, and that the pointy end of a file is called the "tang," the middle of the file is the "belly," and the type of file we really needed to have to sharpen our axes was a "mill bastard." We all liked that name and laughed when we said it out loud (a licence to swear), but not in front of Craig. Axes came in different weights and different shapes, as did their handles, he explained in his monotone teacher's voice. There were full axes, broadaxes, double-bitted axes, half axes, hand axes. "But don't ever pick up a hatchet," he would say, "because that's a dangerous tool. You can swing it in one hand and hit your other hand. You don't want that." Serious business, axes.

Adding to Craig's credibility was the fact that the camp owner and director, Kirk Wipper, who himself was something of a legendary naturalist and woodsman, held Craig in high regard. He would call on Craig at flag raising in the mornings, sometimes to talk about the weather we were likely to experience that day, based on the winds, the air pressure, and other signs from nature, like what the birds were doing. It never occurred to me to wonder how Craig knew all that, but it was clear from these contributions to camp life that he was a student of the woods and all that the woods had to teach. Kirk would call on Craig at mealtime in the dining hall as well, from time to time, to amplify a point about some aspect of camp life or a detail from a particular canoe trip achievement that had been notched by one of the camp's out-trips. Craig seemed to know almost as much as Kirk on many of these topics. But even more than that, on some topics like canoe routes,

Craig seemed to be drawing on a vast repository of knowledge, maybe even bigger than Kirk's.

As a camp director, Kirk had a knack for taking maximum advantage of the leadership potential he saw in his staff. All the CITs would be channelled into a counselling role in their return to camp the next year, but almost immediately, when there were roles to fill in the instructional areas like campcraft, rifle, archery, wildlife, canoeing, or swimming, or in programming duties like organizing all-camp activities and events, he would deploy young leaders with particular skills in those areas. Craig was never the nurturing, warm, fuzzy type of counsellor who naturally coddled or consoled homesick campers. As a woodsmanship technophile and tripper who had a particular affection for maps and canoe routes, his place at Kandalore very quickly became rooted in the canoe tripping program. Craig's forte from the very beginning was researching canoe routes, where other camps had gone, how they got from here to there, and crossing divides between watersheds. Talking to him about canoe routes, even in those early years, was often confusing because the map in his head was far more comprehensive than the map in anyone else's head.

In 1965, when Craig was eighteen and finishing his second go at Grade 13 at York Memorial Collegiate Institute, Kirk alerted him to the growing interest across the country in the upcoming 1967 Canadian Centennial. A group of paddlers from various provinces who were affiliated with the Canadian Camping Association (CCA) were thinking about marking the Centennial with some kind of cross-Canada canoe pageant. Through the CCA and its Ontario affiliate, the Ontario Camping Association, Kirk had been privy to an idea hatched by Gene Rheaume of Flin Flon, Manitoba, to have a race from Edmonton to Montreal using two-person canoes, drawing from the ranks of amateur competitive marathon paddlers. That idea morphed into the Centennial Voyageur Canoe Pageant that, with teams from eight provinces and two territories, got extensive coverage in 1967. But a lesser-known offshoot of Gene Rheaume's idea was to do more or less the same thing, but with trips organized by youth camps. Each camp would take a piece of the route and do its own trip to celebrate the Centennial.

Perhaps understandably, no camp director came forward to offer up a crew to paddle the most dangerous part of the cross-Canada route, the north shore of Lake Superior. As part of his interest in canoe tripping and his ability to see a bargain when he tripped over it, in 1965 Kirk had managed to secure two of the very early canvas-covered wooden prototypes of voyageur canoes made by the Chestnut Canoe Company, which had been deemed not seaworthy enough and set aside for bigger, stronger fibreglass voyageur-style canoes. It was these canoes he had in mind when he offered a crew of Kandalore staff and alumni to plan and execute this difficult journey. And perhaps he did so with confidence knowing the experience, quality, strength, and tenacity of his staff but also knowing that he could assign the route planning, safety, and navigational aspects of the journey to young Craig Macdonald.

So as Craig returned home to begin his first year at the University of Guelph, he was even more enthused about the extracurricular project Kirk had laid at his feet. That winter, Craig reached out to the legendary canoe tripper Eric Morse in Ottawa. Craig and his brother, Rod, communicated with the federal Department of Fisheries with inquiries about information relevant to a canoe trip along the north shore of Lake Superior (through which they learned about an ongoing pesticide project to eradicate lamprey eels from the streams and rivers flowing into the Great Lakes). Craig got deep into the various maps that were available. And when the time came in the late summer of 1967 to actually execute the trip — which was a study in just how many things can go wrong in mounting a canoe trip like this[1] — Craig had a trove of maps in a variety of scales, all carefully marked and annotated. The way he described preparing this cartographic treasure for travel speaks volumes about Craig and his particular affection for maps:

> I proceeded to make a waterproof map canister that would store a very large roll of maps. This was made from a heavy cardboard tube closed on the ends with wood plugs. One plug was permanently held in place with copper nails. I prepared a paper listing of the distances for dozens of way-points for both an inner and outer passage through the islands and across bays. This paper was glued to the outside

> of the tube to facilitate rapid distance calculations. The storage canister was then given several coats of varnish to make it waterproof. While the maps were in use in both canoes, they were carried separate from the canister and protected by placement in a clear plastic bag.[2]

With the Lake Superior trip planning well under way and more or less running in the background, Kirk Wipper had other plans for Craig in the summer of 1966. An old fishing camp on Garden Island in Lady Evelyn Lake in Temagami had come on the market, and Kirk, hungry for an outpost camp to extend the Kandalore canoe tripping program, bought this property and renamed it Outpost Island. Again, the director turned to his senior staff to figure out how this new facility might be programmed, and to Craig, again, to document the permutations and combinations of campsites and canoe trips that could be expedited in the region. This task was the one that connected Craig to Temagami and sowed the seeds for the avocational research that would become his life's work.

So, in the summer of 1966, with groups of senior campers and staff selected to pioneer canoe routes out of Outpost Island, Craig ground-truthed his library and archival research into routes. These included trips up and down the Golden Staircase with stops at Maple Mountain, trips north and west through Florence Lake into the Sturgeon River watershed, and similar explorations to the north and east via the Makobe and Montreal River systems.

Because the camp accessed Outpost Island through Mowat's Landing on the Montreal River, just below the Mattawapika Dam, Craig made it his business to introduce himself to an Anishinaabe trapper called Pete Albany who lived in two canvas wall tents right at Mowat's Landing. And, through conversations with Pete about routes and the Teme-Augama Anishnabai People's deep attachment to this country, Craig was first among the Kandalore crews to learn that Garden Island, the name of which had been set aside by Kirk Wipper in favour of Outpost Island, had beautiful sandy soil and that, indeed, it was the garden of an Anishinaabe Elder and medicine man, Wendawban, who had been forced to move when the waters of Lady Evelyn Lake were raised by the loggers who flooded him out.

In the summer of 1967, Centennial year, while Craig was leading other pioneering trips for Camp Kandalore and preparing for the epic Lake Superior journey, which would not start until late August, I was part of an Intermediate Section cabin group with Counsellor Pat Geale. We were among the first younger campers to venture into Temagami from the new base at Outpost Island. Although Craig was elsewhere, he was very much present in the stories we heard from the resident staff at the island. This included Bill McFarlane, a physical education teacher from North Bay who had been Craig's canoe trip guide in 1960; Peter Anderson, who was a staff member of Craig's vintage; and an old Temagami trapper called Charlie Haultain, an old mining claims inspector who had camped on Garden Island in the 1930s, first engaged by the owners of Garden Island when it was a fishing camp. Charlie had a square transom on his canoe that he swore to us was installed because he'd once sliced a slab of bacon on the stern while his boat was upturned at a campsite. Later, a bear came along and ate the end off his canoe, with its lovely bacon grease garnish.

Their own stories notwithstanding, it seemed that everybody on staff at Outpost Island referenced Craig in one way or another. His influence on the growth and evolution of the Kandalore canoe tripping program and the related trust that Kirk had in him were both clearly evident as we went about activities at the base camp and multi-day overnight out-trips that included climbing and descending the hairy goat-track portages of Temagami's Golden Staircase, like Fatman's Squeeze, and climbing the fire tower on top of Maple Mountain.

The official Centennial Voyageur Canoe Pageant was in full swing that summer, between its start at Rocky Mountain House on May 24 and its exciting conclusion with a last sprint race at Expo 67 in Montreal. This colourful flotilla, with its brawny participants, gathered much media attention along the way. So much so that the Canadian Camping Association's cross-Canada canoeing initiative, which was perhaps even more logistically challenging for its organizers and heroic for its young participants, basked in almost total obscurity, at least in the public eye. For many, the first time they heard of Kandalore's epic voyageur canoe trip from Thunder Bay to Michipicoten (they didn't make it quite to the

Sault, their original destination) was in Craig's long article in *Nastawgan*, published fifty years later!

In the summers of 1968 and 1969, Craig was leading long pioneering trips for Kirk Wipper. He'd become something of a shadowy presence in the camp itself for campers like me because there was always so much going on and he was always on the trail. The trips he led with senior campers often had formal send-offs and welcomes on their return, but, it being a camp, the tanned faces at the centre of these celebrations would be the campers, not so much the counsellors and certainly not the guides, like Craig, who, if he was there at all, would always be lurking in the shadows. His job, as he saw it, was to pick the routes, to find ways to get groups through the various challenges these northern rivers presented, and to return them safely to camp and to their parents back in southern cities. Craig never took joy in accolades of this or any kind. Even when Kirk called him out for his exemplary service, Craig would shyly look at his feet, blush, and back farther into the shadows, if that was even possible.

In 1970, however, my last year as a camper before joining the Kandalore staff — Craig was 24 and I was 15 — Craig was assigned the job of guiding that year's exploratory trip into the wilds of western Quebec. By now, the roll of maps that he had gathered for his 1967 trip was buttressed by sheaves of other topos and by conversations with a growing number of Indigenous people, like Pete Albany at Mowat's Landing, about how Indigenous travellers moved through all of this country. We started the trip at the historic logging town of Temiskaming and moved up through the Kipawa River system to big Lac Dumoine. From there we looped back down to the Ottawa River via the storied Dumoine River. This route had been taken years before by the more ambitious American canoe tripping camps from Temagami, like Keewaydin and Wabun, but for Kandalore it was new ground. And leading us there was Craig Macdonald.

For me, the trip was hugely influential. Through the '60s, as I had moved from day hikes and cookouts to overnight canoe trips and on to multi-day adventures in more remote wilderness, the Kandalore way was to portage around any significant whitewater, whether we were going upstream or down. But to prepare for the Kipawa-Dumoine trip, it was acknowledged that we had graduated to a real river trip, which meant we needed to learn a new set of skills. We went to the outflow of the dam at Twelve Mile Lake, just down the road from the camp in Haliburton, where we learned some very basic whitewater canoeing strokes, like braces and crossdraws, and manoeuvres like ferries and halfback and fullback turns to get in and out of back eddies.

From there, in the same sixteen-foot fibreglass canoes we would use on the actual trip — three to a canoe, no flotation devices — we moved to the Gull River, just below where it flowed through the dam at Horseshoe Lake (now the Minden Whitewater Preserve), to string together whole routes to learn the "tongues" and "haystacks" that make up the essential character of moving water and to develop the skill and respect necessary to safely negotiate a rapid. This was relatively new for everybody at Kandalore, including Craig and the other staff members who would be on the trip. But, to the best of my knowledge, Craig was never at these practice sessions.

The trip was just shy of three weeks long, which was an eternity to youth like me who had only done, by that time, trips of maybe a week or ten days' duration. As I look back on the experience as a whole, there are peak moments I remember — more highs than lows, but certainly a few of both. But the memories are more impressions and sensations than notes, photographs, or marks on a map. Even after I'd gone back to the Dumoine repeatedly over many years as an adult, as a guide for Black Feather wilderness adventures, or just for fun — a full Thanksgiving dinner on Lac Benoît flashes to mind — that journey of 1970 remains a happily hazy narrative thread that stitches together moments of terror and exhilaration and the sensations I can still feel to this day but not adequately describe.

Having spent six days paddling *up* the Kipawa River, we finally crossed big Lac Dumoine and started down to the Ottawa. I can still feel the rush of powering across a raging current into a back eddy flowing the opposite direction with equal force and, as a well-practised team of three in a heavily loaded canoe, leaning into that roller-coaster 180-degree turn that happened in less than a heartbeat. And in the moment of that indelible wet and wild realization of our accomplishment, my heart pounding, my hands shaking, Craig Macdonald's measured cry to alert everybody that the canoe right behind us had flipped on the same obstacle and strewn people and packs into the welter of unforgiving whitewater.

Craig was seething that the two heavy food packs in that canoe, one of which contained the main cook set and half the billy tins, had gone to the bottom of the Dumoine River, never to be seen again. But, always the teacher, this was a chance for him to remind us that we always had to build in redundancy to our packing, for this exact kind of snafu. In this case, we had packed a second pot set and the other half of the billy tins in a separate pack that was carried in a different canoe, and they would do us for the remainder of the trip. We would have to eat in shifts. *C'est la vie.* I also remember that Craig had no sympathy whatsoever for the plight of his assistant guide, who lost his only supply of cigarettes in the mishap. "Those dang things'll kill you anyway. Might be a good time to quit," he scoffed with a satisfied little grin.

A half century later, I was having conversations with Craig to tell his story — this story — and, because we shared that Kipawa-Dumoine trip in the summer of 1970, I thought this might be a chance to get inside Craig's memory, to see how it works, how he connects things in space and time, and how he carries that information forward into new experiences, new ventures. So on Thursday, July 16, 2020, almost fifty years to the day since we'd spilled out of a smelly school bus in Temiskaming after the long drive from Haliburton, we sat down in his house in the little town of Dwight, Ontario, just west of Algonquin Park, and revisited that trip from so many years before. It was an illuminating glimpse into a remarkable mind and memory. What follows is Craig's voice, unless otherwise indicated. Keep in mind that these recollections are drawn through a lifetime of other rich and intense experiences.

> JAMES: What do you remember of our 1970 Dumoine River run?
>
> CRAIG: I won't bore you with the start of the trip. But I'll pick it up when we went into Quebec. We went to the CIP [Canadian International Paper Company] mill and we had a tour and we saw the digestors and the great big

woodpiles outside with the conveyor belt taking them up to the top. I had organized it slightly ahead, just a little bit of phone calling, arranged for the tour. And they were happy to give us a tour. We saw basically the whole plant operation, including the big spools of pulp, we saw them cut one off and then pile it with the other spools on the floor there.

Then we went up through the town of Temiskaming, and we went around the circle and saw the statue of the cupid (or whoever it was) peeing in the water. Years later, Margaret Atwood had a book called *Surfacing* and she made fun of this little fountain in the middle of town with a peeing individual. We circled through the town and then we went up to Letang and we saw the river roaring down through there. We also saw the flume, big flume that ran down beside the river there toward Lake Temiskaming.

The thing that got me the most when we were driving up there, two things. One was we went past Letang there and there was a residence that was on virtually a sandpile and there was no vegetation there whatsoever and out front there was a sign that said "El Rancho." What the fellow had done was take three or four cedar trees and turned them into saguaro cactus. He had taken spikes and nailed them into the tree like the thorns on the saguaro cactus in the desert. It was pretty humorous.

The other thing that impressed me was that as we went up the road to Kipawa, there was a Native crew of men working on the right-hand side. What impressed me is that they had no power brush saws. They were all using loppers. The work of clearing the brush on that right-of-way by the road, on the right, was being done by hand. Long-handled pruners. I thought, God, this is the epitome of a make-work program.

I remember coming up to Kipawa and I think we talked to the ranger. If you give me a second, I could probably remember his name. We started off on the lake. Of course, we had a lot of packs. And it was a very great concern because Kipawa, these lakes we were going to be travelling were big lakes, and we were way overloaded. We had fairly big guys. These were not small campers. These were, what would you call them, sub-adults? So we were heavily loaded and that was a great concern....

I asked [assistant guide] Doug Wipper [Kirk's oldest son] to take additional packs and I took additional packs. As a result, we were incredibly low to the water and just wallowing up the lake. You had to be very, very careful where you were with regard to your balance, or you could easily roll the canoe. Because those canoes were not designed for that type of load. So we went up and we went through a narrows. Everything was okay, we had a bit of a tailwind. It was okay.

But then, we came up to Corbeau Point, across from Corbeau Island, on Lac Kipawa; the wind started to pick up, because it was getting later in the day. That was a concern. Both Doug and I were taking in water. Because we were so low and we were having to deal with these waves. And some of the canoes started to box us in to break the waves from coming from the back. And that helped. But it wasn't enough. And I remember my canoe was okay. We sort of made it through this. But Doug had to go to shore. He was literally swamping. Doug had to dump the canoe, just after we rounded the point....

Then we went to Grindstone Lake. And there's an island there where Native people used to take their axes. Why it was called "grindstone" is ... they had ... the rock is very, very fine texture. And there's a rock there you can take your axe to it, rub your axe on the rock and sharpen it. Usually, the rock around here has larger flakes, the pegmatites and all that stuff, the crystals are too big. We didn't stop there. We went into Grindstone and then made the big turn and went up into McLachlin.

We're going upriver, but because it's all flooded back, the first portage that we did was mercifully a good portage. It was on the right. And it was sandy. And they also had built a tramway so we had a narrow gauge railway tracks going up there. Remember that?

JAMES: No, no recollection of that.

CRAIG: We didn't put our canoes on the cart. We portaged beside the railway track, the little tram. We had one counsellor that was shaky. But most of the kids were experienced canoe trippers. Charlie was his name. He was a nice guy. A sort of average height. He hadn't previous camp experience. This was his first time at Kandalore. You'll have to remember, because I've been on so many of these. Like you were sterning your canoe? Right? You were what we called the "kiddie canoe." Your canoe had a bad gunwale that we later had to fix. And I think you were the only kiddie canoe. And then we had this Charlie guy. And then there was Doug, and myself, and Pat Geale. We had six canoes....

The weather was kind of deteriorating. We went up into Sun Lake and then I could see a lot of mare's tails and such. It just got worse and worse until we knew we were really going to get hit. You just had this feeling something big was coming. It hit at Lac Sairs. And then we put in on the left. I can remember it was a sandy campsite. But the sand was rough because it rained hard for three days. My concern was what are we going to do for firewood? Because we're in this poplar grove. I was able to go back and there were these poplar chicots [dead poplar trees].

And, of course, I had a bowsaw with me, so I could saw them out. So we did quite well. We were able to have good fires for our meals but the weather was just the cruds. It rained hard, hard in the night. Got up in the morning and the clouds were barely over the trees and it was blowing east. It was still in the thick of things. We're going to take a rest here. So we basically stayed in the tents because it was raining that much, and hard. And then, when we got up on the third day, it was still raining. But the wind had shifted. So we knew the weather was going to break, so we said, "Okay, let's get going."

Our first serious portage was Elliott Falls. We were going upriver and, of course, the river was incredibly high. The water was right into the trees, into the alders. We were working on strong currents. Heavily loaded. We just didn't have enough oomph, with the loads we had on, to barge right up the middle of the river. We had to use every trick in the book. We went right to the shore to get the slower water. I remember some of the rips were so tough that we were hand-over-handing on the alders in one spot.

We were way too loaded to have poles. The canoes were too small. The shoreline was hostile. You couldn't even line the thing. One of the disappointments on that trip was the equipment that we used for lining. I said to Kirk, "We're not going to do this trip without lining ropes. There's no way you're going to ask me to go all down through here without lining ropes." So what they did, is they went into Minden or someplace and bought us these sisal ropes, like manila ropes. And the doggone things, unless they were wet, gave you splinters in your hands. Remember that?

JAMES: No.

CRAIG: We got a bale of it and cut it up. We had to have six lining ropes that we had to use. And I think they were all about thirty feet. We didn't have ropes on either end, because we didn't have enough rope. So we just did the single rope line. Which worked out fine. I was used to that sort of thing going upstream, you go light. You put one guy in the canoe and he paddles on the bankside and you put a rope around the bow seat, not underneath, although I think we did make harnesses underneath going downriver. Nowadays I don't do that....

The big rule that we followed on that trip was that you don't line above the knees. You always want to keep in shallow water. But that was really hard going up the Kipawa. The high water and the hostile steep-sided banks. At one point I was chest deep in water and I thought I was going to get into trouble. We did not have any PFDs [personal flotation devices] with us.

I remember, going downriver, I said, "Do not go into the water above your knees." There's just too much risk of being swept off and having an accident. You lose control of the canoe or you get hurt. The manila ropes, we deliberately soaked them to try to get rid of the splinters, soften them up. Going right into the hands. It was awful. In the old days, the voyageurs had quality rope. They had good lines. Strong lines, but not thick lines. You want them thinner so they don't catch on the water and do things that you don't want the line to do.

So, we went up to Elliott. That was the first big one. And we continued up through very crappy weather. We got a little bit of a break when we got up to Wolf Lake. That's where those fishermen were. And there was an old Native community. There was a dam. We camped on the left. And the old community was on the right. You could see where the abandoned buildings were where the Native people lived at Wolf Lake.

I remember, we got dinner over in fairly good order. Then, after dinner, we had the tents pitched and everything and I said that I want to go out and try out some of these whitewater manoeuvres. So I went out and did a couple. And then Doug Wipper went out and did some too. By that time, I'd parked the canoe and I was up on the dam. And Doug paddles out, by himself, but I'd beached my canoe and then gone up on the dam. And then I was sort of watching, spotting from the dam at the outlet of Wolf Lake. One of the scariest moments of my life happened right there.

There was a pier in the middle of the dam, so there were two separate rips coming down from the lake. Huge volume of water. It's flood conditions. He went out into the current and immediately snapped around and sunk the canoe. He disappeared. And so did the canoe. And he never came up.

So I ran off the dam and jumped in the canoe and paddled out downstream of all of this and … I was almost to him and he surfaced, just bursting for breath. He was down there long enough for me to run off the dam, get in the canoe, and paddle out. So how long he was under, I swear he was down at least a minute or more. Like a long, long time. The only reason he was able to do that was he was a licensed scuba diver. He was a commercial diver. So he could hold his breath and he didn't panic. But he damn near drowned.

He came up and I got him in the canoe. And after he was in the canoe, his canoe came floating up from the bottom. It surfaced. And I said, "Holy shit!" I'd never seen anything like that. I didn't even know it was possible. I couldn't even see the feature that sucked him down, and held him down there. I asked him what it was like down there? And he said, "Oh I was tumbling all over the place and I have no idea because it was all froth and bubbles and stuff. I really don't know. I just know that eventually I got out of it and came up." I never did figure it out.

The interesting thing about Wolf Lake was that I had been told that that's where the woodland caribou were. They've moved out of Temagami and all the rest of Kipawa but there was still a resident herd at the east end of Wolf Lake. But the land around there is so low that the end of the lake was below the horizon. I don't know if you remember that.

JAMES: No. No recollection of caribou....

CRAIG: We went up into Lac Watson and we worked our way up to Dumoine and we camped on the northwest side of Quabie Island. I don't know if you remember that. There was a fish drying rack there. There were four poles, like a pyramid with the cross poles and the racks for drying fish. I thought, this must be Native people doing this. I subsequently found out who Quabie was. He was a Native person that lived there.

And then I remember, going out with somebody, I don't think it was Tom Darmour. I'd taken off my shoes and socks and we paddled around Quabie Island and I remember coming back. Stable flies biting my ankles. So I'm doing a bit of a dance in the canoe to keep the darned things off me. And I'd never experienced that before. Those biting little flies are nasty buggers.

Out on the water, it was awful ...

I was very relieved at Quabie Island. We'd been doing a lot of paddling in seriously high water. Gone through a lot of lakes and our weight, fortunately, was getting down. So we were much more capable in the water. But even still, those canoes were just not designed for whitewater with three guys in them. You couldn't. The only thing we could do. You couldn't backwater. You couldn't do that sort of thing. We had a keel on. We were deeply in the water. But we'd made it to the height of land.

We didn't have trip reports from Keewaydin or Wabun or any other camps that had been down the river, but I knew those guys had done it. And, of course, you'll remember when we started down the river and did one of those early portages that we found one of their wrecked canoes with the “K” carved in it. They actually cut the square of canvas where the “K” was on the bow, so the “K” had been removed but the knife marks were still in the wooden planking. We knew it was Keewaydin. You’ll remember it was on the right side downriver....

Below there there’s a rapids and then there’s a waterfall and there’s two short portages. You can run the rapids and then you have to portage the waterfall, what do they call it? Bear Rapids. *Mukwa boweetig.* I said, I don’t care, we’re not going to start with that type of thing [crashing a canoe in the rapids]. We’ll portage the whole thing. We took the long portage. We’re not ready for this stuff. Early on, when Doug wiped out and we lost the food pack, we realized the canoes just weren’t that manoeuvrable. Basically, all we could do was draw and cross draw in the bow to get things around. It was running slightly faster than the water, maybe neutral, and hoping for the best. No backpaddling that I recall. No back ferries. But that’s what we did....

I recollect just about every campsite on that trip, like every trip. One that sort of intrigued me was on what they call “Little Italy” on Lac Benoît. We were out on the bay, camped there. That night we had an unusual number of cans for some reason. I remember in the morning pounding them out. We weren’t leaving garbage. What we were doing was burning the cans out in the fire, flattening them and then carrying them out in plastic bags. So, by the end of it, we had a pack of cans. At least half a pack of cans, a large number of cans. I remember that.

We ran what we could but then there was some stuff that I said, “Nah, we’re going to line some of that.” So we lined a bit. I remember lining on the west side and I also remember lining on the east side. And one of my recollections of lining on the east side was, there was a bunch of hazel brush that was along the shore. Beaked hazel, like the stuff out here, grabbing that with my bare hands and getting all those fine hairs in your finger joints. Uncomfortable.

Yeah, there was one rapid there, that sticks in my mind. I didn't actually see it because I was ahead. There were about six guys that saw it, a rock statue of a face looking upstream. We were lining on the right-hand side, on the west side. I remember it was so flooded and there was a campsite at the start of the portage. The poles on the campsite from Keewaydin were leaning on a tree and they were fully in the water. The whole campsite was under water. That's the kind of water we were dealing with.

It was very, very high water. It was 1970. They started recording the levels in '47 or '48, something like that. So, it was somewhere before that flooded campsite, so we were lining on the west side. It was like a face, facing upstream, that had been carved into the rock. So it was just the face, not the body or anything else. And it was stuck in the rapids and it wasn't too far out from where we were lining. It was rock. Somebody had gone at it with a chisel. It wasn't a quirk of nature. It was obviously carved, looking upstream. If nobody is able to find it, my suggestion is that some tree has come down and pushed it over. Hopefully it is still face up. The six of them were quite impressed. That's all I know....

So we keep going down. With lighter packs. We're getting better at the whitewater. We're cooking with backup pots at that point, as I recall, because Doug had lost one set when he dumped. Billy tins went missing. It was a kitchen pack that had gone astray, gone to the bottom.

Pat was the kid manager, and the disciplinarian. I didn't get too involved with that sort of thing, people messing around in canoes, or messing around with an axe. If somebody got into something like that where they might get hurt, obviously, I'd be right on it. But in terms of squabbles amongst kids, Pat was the man. He was the counsellor. The division of responsibilities was very clear. I was looking after the route and overall safety. That was my role....

The one thing that was a bit troubling, going down, was your canoe. I was afraid that if we didn't get that gunwale fixed up you were going to have a serious rip in the fibreglass. So, when we got to Sheerway, we camped downriver and I asked Doug to go back up to the hunt club to see if he might borrow some tools. "We've got to do something about this canoe," I said. It's way beyond what it should be. It needs immediate attention. We've got to do something with the gunwale. He was very fortunate that the caretaker was there. The caretaker had tools and a spare gunwale wood, so he was able to take your gunwale off and put a new gunwale on. That saved the day. The thing is, that canoe should never have gone out....

It was very apparent in the rapids, those canoes were too small. We had to be very conservative about what we were running because we couldn't turn the dang things. They were so sluggish. You had the bowman work like heck to move it and, boy, if he didn't start moving early enough you'd be onto a rock. There was a couple of dumps. That Charlie guy, he dumped once. The guys were too big for three to a canoe. They just didn't feel stable. They were tippy and just didn't seem right for what we were up against. I said that if we're to have a future in this stuff, we have to go big. So we did. And it worked out.

By the time we got to Grande Chute, I was feeling a bit more relaxed about the situation because I knew it was nearly over. We had Red Pine Rapids below. But my job was getting pretty well done. So we went down to Red Pine and that was okay. And then we went by that huge rock face, just at the confluence of the Fildegrand. An amazing piece of rock. That was spectacular. I really enjoyed seeing that. I didn't know that was there....

We did camp at that campsite on the east side near the bottom of the river, just below the last drop. And I remember it was pretty darn good fishing there. Tom Darmour was fishing and he caught a beautiful bass. Then, we had a day or so to kill before we were to be picked up at Stonecliffe, so we had some options. It's kind of interesting because ... do you remember going up the fire tower at the mouth of the river that was run by Ontario Lands and Forests? Right at the mouth of the river, on the upstream side? We walked up the trail and up to the tower.

JAMES: No. I have no recollection of that at all.

What I did remember, sitting there in Dwight listening to Craig's recollections of the trip, was something he was a little hazy on himself that had to do with some fun we campers had at his expense at the end of the trip.

Tom Darmour, the kid who caught the fish at the last rapid on the Dumoine that so impressed Craig, was a seriously overstuffed kid from Cleveland, Ohio. Toad, as he was called, had a great sense of humour and was forever needling the trip staff, Craig in particular. At one point — as is *always* the case on a canoe trip — campfire conversation came around to food. Favourite foods. Foods you miss. Foods you're glad to be without. Food. A very popular canoe trip topic for everybody but Craig.

Craig would sit and eat his meals in silence, often on his own. But once, egged on by another kid from Ohio who was on the trip, Toad mentioned in passing that he had a particular affection for McDonald's Big Mac hamburgers. He said that on more than one occasion at home he'd scarfed ten Big Macs in a sitting. That claim was enough to break Craig out of his reverie.

"There's no doggone way that a human stomach can hold ten Big Macs," he said.

Toad replied, "Oh, yes it can. With fries."

And with that Craig, so uncharacteristically kibitzing with the campers, which was *never* his thing, turned to Toad and said, "There's a little greasy burger stand in Stonecliffe. I'm thinking if we have time tomorrow, after we visit the Des Joachims power dam, we should paddle across the river and have a little bet. I bet that you can't eat ten hamburgers. You can pay for the first two, because that's what a normal person would eat. But if you can eat another eight, all dressed, I'll take a loan from Kirk's emergency money, and pay for them. If you get stuck, you're on the hook for the bill for however many you shove into your mouth."

"You're on," said Toad.

And with that, others got into the action, just to bug Craig, saying, "Hey, Toad, you eat those ten hamburgers, I'll buy you a strawberry milkshake." And on it went, the hype building minute by minute until we all beached canoes on the Ontario side of the Ottawa River and hiked up and over the railway to the highway and along to the Stonecliffe burger bar.

We explained what was going on. The proprietor was intrigued by the wager and got to work at the grill. Craig at this point was feeling quite confident. "Make 'em extra greasy," he yelled from the parking lot as Toad and his entourage settled in at a bunch of picnic tables.

Good to his word, and to everyone's astonishment, especially Craig's, Toad stuffed in the first six burgers two at a time without breaking a sweat and then took a short break to suck back a strawberry milkshake and crunch on a few onion rings. The final four all-dressed burgers went down one at a time, the last couple with just a hint of noticeable effort. Then Toad sat like Henry VIII with his entourage hovering around, snacking on an order of fries and smiling like a Cheshire cat. All Craig could do was pay the bill and say it was time to head back to the campsite for dinner.

It was about a half hour's paddle back across the river, during which there was much taunting of Craig, who, while feigning disgust, appeared to be enjoying a bit of adolescent fun. To show that there was indeed additional stomach capacity in this championship glutton's belly, our friend Toad ate a whole roll of Tums, like candy, popping one into his mouth each time he'd catch Craig's eye on the way back to the campsite. But it was the image of Toad, parked in the middle of a canoe, like a big fat ... well ... toad, being paddled by his two canoe-mates, that Craig's mention of his name brought flooding back as we relived this long-ago trip.

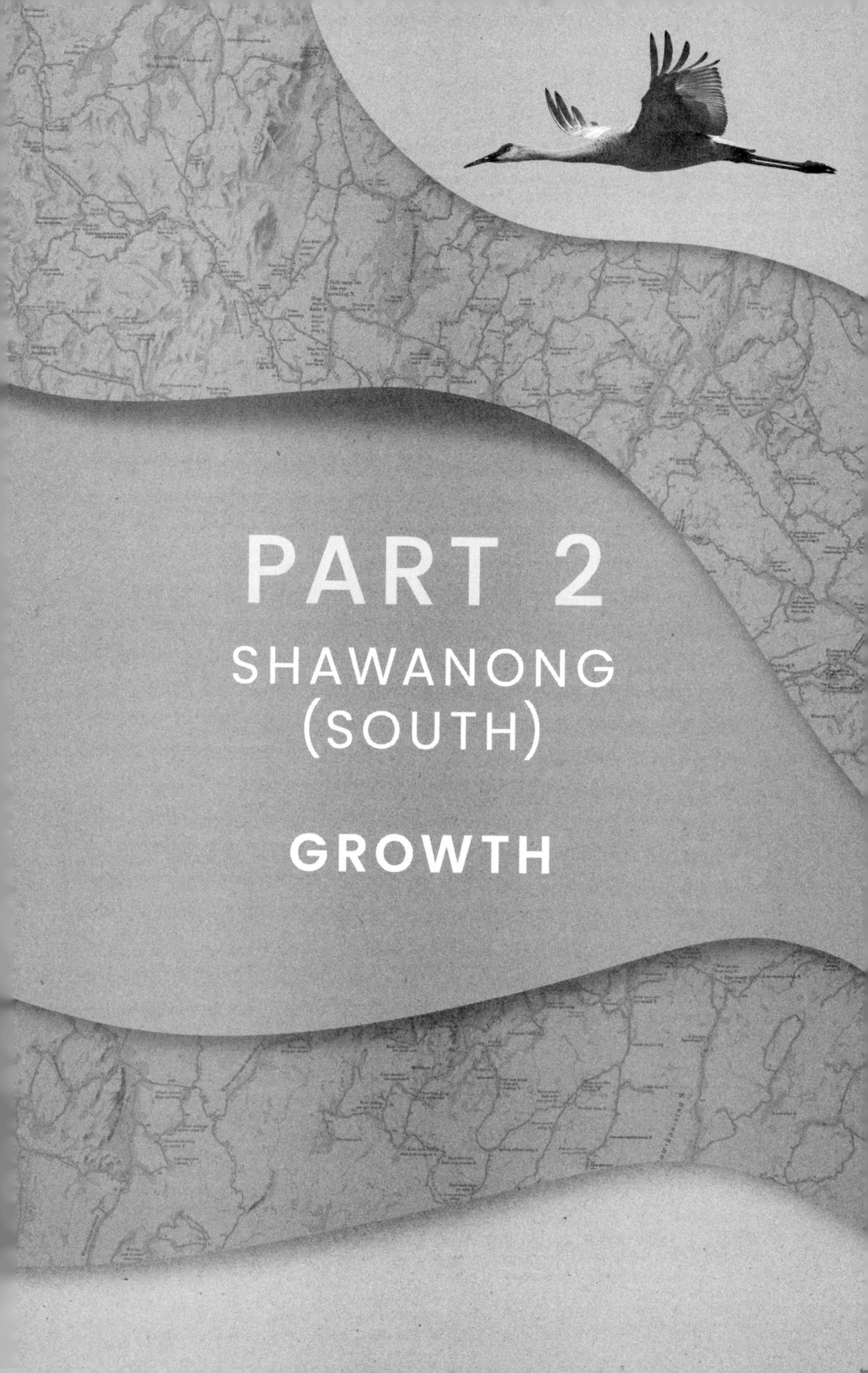

PART 2

SHAWANONG (SOUTH)

GROWTH

Mr. Macdonald Continues Testimony in the Supreme Court of Ontario

THURSDAY, JANUARY 27, 1983, 11:45 A.M.

Mr. Clark (Lawyer for the Defendants) Q. Now, [your map] doesn't show the whole of nDaki-menan, does it?

A. No, it does not.

Q. As to the portion of nDaki-menan which it doesn't show, have you gone through the same process accumulating data, as with respect to the region which it does show?

A. Yes, I have.

Q. And the area of nDaki-menan which isn't shown, does that relate to the northern portion of nDaki-menan?

A. Yes.

Q. So then, the field work for the whole of nDaki-menan has been completed, but if I understand you correctly, that field work has not yet been all summarized on one single map?

A. That is correct.

Q. Is there any particular reason why?

A. No reason at all. I just decided to do the south first and this, maybe it should be emphasized, this is not a territorial map, this is not coterminous with any territorial boundaries. What this is, is a geographical guideline for Quebec and Ontario and running west as a part of the research I have been doing right through Western Quebec, Northern Ontario, right across the country.

Q. Could you just as easily have started with the summary as to the area immediately to the north, as to the area which you did start with?

A. Yes.

Q. Is the quality of the information the same with respect to that?

A. Yes, it is.

Q. Now, would you please explain in turn the Nastawgan shown on this exhibit or some of the Nastawgan?

A. Yes, okay. I think probably the best way to tackle this is to discuss simultaneously when I discuss the routes, to discuss the geography. If you don't have a handle on the geography, we are going to get terribly lost and it is going to be very difficult.

His Lordship: I am going to have difficulty following this from a distance. It is all white with some black and grey marks, nothing distinguishable from this distance. When you do something could you expressly describe it or use coloured pencils to point out what you think is important? It may be if I was much closer I could read something on it, but from here I can't read a thing on it.

The Witness: What I think the important thing would be for us to do here, is just to say, for example, if I intend to start from this corner, as long as you know we are talking about this general area.

His Lordship: From here, I don't know whether that corner is Sault Ste. Marie or Quebec City or North Bay or Bear Island or where it is.

The Witness: We will have to get it closer ...

His Lordship: Until I get within two feet of it, I can't read the names.

The Witness: What I suggest, I will give the name in Anishnabai of what I am talking about, where it does have an English name, many of these do not have, they have no name whatsoever, I will describe it as being close to a known lake.

His Lordship: Doing it that way may well be the easiest.

Mr. Clark: Perhaps you could start by writing with a red pencil or something and a line where you are going.

The Witness: I will number them. The first lake we will discuss is this lake here and the name is Ka-addickumaigonse-ka sagihaygun. What that means —

His Lordship: Have you marked it?

The Witness: Yes, I have called it number 1.

His Lordship: It is in the left upper central portion of the map.

The Witness: Yes. What I am going to do is I am going to give you the native translation, you may find it better. I could give you the English name for the lake so there is no question what I am talking about later on.

His Lordship: I think you better give the English names. All the other exhibits have English names on them.

The Witness: I am using the current name. The current lake name here.

Mr. Clark: Where the Indian name is relevant to show ethnic identity or culturalness, it might be better to give both in those cases, I think.

The Witness: Okay. I will do that. Stull Lake is the current name. Stull was a surveyor in the 1800's.

His Lordship: I don't think we need all the historic evidence. Your principal evidence is going to be where the various routes are and perhaps linguistic aspects of it. There are no exhibits filed that have the routes mentioned and the Indian names you are using and if you could relate it to what the English names are wherever possible.

The Witness: Okay. What the Indian word means is herring lake. Herring is a small fresh-water fish that often travels in schools surfacing en masse and they ripple the water, probably everybody has seen them. This lake was the site of a very old settlement of Teme-augama Anishnabeg on the north shore towards the east. The late Donald McKenzie told me there was an Indian cemetery located near a settlement on this lake.

In the surveyor's report for the survey and exploration of Northern Ontario in 1900, this report notes the remains of an old trading post and Indian clearing at this location. The specific page reference [in the surveyor's report] is page 104.

This will be submitted under another topic, this specific page. The herring, whitefish and lake trout of this lake were netted in the fall and provided an important food source over winter. This fish is also an important food fish for the lake trout. The lake trout feed on this fish. According to Donald McKenzie, he believes Ka-addickumaigonse-ka sagihaygun is the name of this lake and is the uppermost lake in the Sturgeon River chain with lake trout.

This lake is on the crossroads of an important interior canoe route that connects the headwaters of the Sturgeon with the Montreal River. On a broader perspective, it is one of the connections between Lake Nipissing and James Bay, in other words, by going right by this place, you are on a route that will connect you with major geographic features in Canada.

Unfortunately, I don't have any population counts for the community, but I can say it was definitely log structures. It was not tipis.

His Lordship: That is the old settlement?

The Witness: Yes. As I just mentioned, it makes mention of the fact in the 1900 report that there were remains there that they observed of this place.

The second location that I want to mention is one that I am not particularly sure about. I have difficulty with this one. It is called, I believe, Addumaig Sahihaygun. Hamlow Lake is the current name.

The Indian translation for that is Whitefish Lake. It is directly below, south of Stull Lake. The word Sahihaygun, in connection with the last one, the language is irregular. Sometime you need this word and sometimes you don't, to describe lakes. The word itself means "lake."

Usually if you have a locative case ending which means a case ending with *ng, ang, ony, ing*, you do not have to use this word but often, if I said Pine Lake, Chingwok Sahihaygun, White Pine Lake, I will have this word often and I think if you could, for the purposes of recording, you could put "s" meaning lake at the end of the others and when we put "s" it will stand for lake.

I know when in some certain cases it will be required. The next one I would like to mention —

His Lordship: Is this number 2?

The Witness: Yes, I have placed the number 2 on the map. Now I am going to number 3....[1]

Number 9 is Wabi-sheep Sahigaygun. The current English name is Staouffer [Stouffer] Lake. The Indian Name is White Duck Lake. I am going to describe these routes in a little more detail [because] it is important to establish the geography before I get into them in detail here. The river is very rocky, as I have said, and you get down to a place, a river that flows in from the north and its name is Pilgrim Creek.

His Lordship: Is that the Indian name as well?

The Witness: No, you are not going to believe this name, brace yourself. Mo-wbo Zibi. That means liquid shit river or shit liquid. Before we get further into this, I should say that traditionally in the Anishnabeg language, there was no swearing. These are not considered — you couldn't curse, you couldn't swear. The worst you could call a person was the devil or something like that. These are not considered crude....[2]

The next one of major importance or significance is as we are working eastward here and its name is Shabomin Sahigaygun (Number 54). What that translates to is Gooseberry Lake. This again is another trapping centre and in recent history, it was the site of Larry Turner's camp. He had a trap camp out on the islands on this lake.

Now, just to give you an idea of the date, I will tell you the story and possibly you can tell me the date, because I don't know the date. I know it is before the Second World War, but one of the fellows, Larry is now dead, as far as I know. One of the fellows that accompanied him on one trip up further north tells me he went to a tremendous amount of work to haul in a bunch of wet cell batteries, a way, way, back in the bush and got this radio set up and it was the first time they were ever going to have communication with the outside world by radio.

This was a new thing. They turned the station on and it was Orson Welles' attack on New York City, so they had no way of knowing what was going on in the rest of the world except that New York was being attacked in the outside world and then the radio broke down halfway through this and they had no idea what had happened to the rest of the world and were wondering whether it was even worth living or bringing out furs when they were coming out.

I don't know, it must have been '38, '39, somewhere in there. I will mark that down in green. Actually, maybe I should make a side note here, there is a route up to the north to the place called Regan Lake, also a trap campsite. I do not know the Anishnabai name for this lake....[3]

Number 72, Abi-jibee S. Again, this is one of these that is not a traditional name, just what it is called now.... What it means is Charlie Moore Lake, after Charlie Moore, Senior ... who trapped this area and even despite a heart condition in his later years, he went out and trapped that area and he was alone at the time and collapsed out on the lake and died in 1958.

This is the story of how [his son] Sonny Moore went overland, because there was slush on the lakes and they couldn't land an aircraft to investigate his disappearance, and he found him on the lake and he used a sleigh, I guess, or I believe it was a hand sleigh, to haul him back to Temagami and he apparently did this without stopping. This is an amazing feat as you are talking about something almost unbelievable — it is a physical feat that not many people could even conceive of doing. I couldn't give the equivalent of labour involved in doing something like that, but anyway that is the story....[4]

Number 81, Ojeeg Pukwudina. This is north of Pinetorch and there is no English name. What it means in Anishnabai is Fisher Mountain. The Ontario Forestry Branch established a fire tower on this mountain and why it gets its name is, and probably still is, but a long time ago there used to be a fisher run on the mountain.

That is not to say it would entail going over the mountain, but where the fisher run is, there is a corridor which the animals come through. Of course the fisher is a very important animal to be trapped and this corridor may be an eighth of a mile wide and the fisher will run to quite a distance and circle around and come back. They may do it on ten to twelve day cycles. This would be a place you could track them.

I guess it would be in the thirties, this animal became so valuable it was profitable and if you saw fisher tracks and it looked like it wasn't going to snow, you would actually track the animals and people from this Fisher

Mountain actually walked entire townships tracking these animals down.

When they found them, they would track them into a tree, hollow log or something like that, and gradually be able to shoot them or they would climb something like a yellow birch tree and ... their favourite habit is to go into the hole of the tree and you would light a fire in the tree and smoke them out and the fisher didn't like the smoke and they would go out onto a branch and they would be shot with a .22 or something like that.

Fisher pelts back in those times were getting one hundred dollars a pelt for them and you are looking at something extremely valuable. One hundred dollars would go a long way. Two or three fisher a year and that would set you up for the winter. That is what that is all about.

The next one here, you are going to have to brace yourselves, this is not intended to be offensive. It is humourous, but not offensive.

Number 82, Mo S. There is no English name. What it translates to is Shit Lake. Number 83, Ogee-dagewan Mo S. That means Upper Shit Lake. These lie to the north of Fisher Mountain....[5]

• • •

Friday, January 28, 1983

Number 149A, Toong-dee-ay Zibi. Again, this is not offensive, it comes out bad in English, it is humourous, but not offensive. What this is, before I tell you the exact translation, is a canoeing term that refers to sitting up on the thwart, being the crossbar in the canoe and bending over and having to do this with the canoe blade going through a very narrow creek overhung with alders. The canoeing term if translated freely comes out as open ass river, a technical canoeing term....[6]

The [Anishnabai] language is so much better than our own, often they can say in a single word what would take a whole sentence to convey to us in English, and then not satisfactorily. However, it should be said also as far as modern technology, the language is hopelessly cumbersome. For example, talking about the word for "pie." In Anishnabai they originally, now they have a different word, but originally they would have to say the word for pie is ishkotay-odawban-mijun. What that translates out to exactly, and this is an illustrative case, "fire sleigh food." The reason why it has this word is this is probably the first time they saw pie was on the trains when they were built by the CPR. That is a very difficult concept to develop the word. They didn't have anything for it. They came up with this, they had no word for locomotive or train, and they invented the concept of fire sleigh. It works both ways.

His Lordship: Mr. Clark, I wonder what significance this detailed name is other than the general statement that there are many words that can't be exactly translated and other words such as snow, there may be many different words in the Anishnabai, depending on the conditions. Why get into a particular word like pie or any other particular word? I am having difficulty following this.

Mr. Clark: When you get words like annuity and money, perhaps it is the most critical issue in the case, and I am setting it up.

His Lordship: Well, I can recognize that this gentleman is very knowledgeable in many words.... If he is qualified as a linguist, and I am not saying he isn't, but from what he has said just strikes me that with his qualifications, I think it is unnecessary to go into a lot of these details which I don't follow.

Mr. Clark: My lord, I don't follow you at all. This is a man that is proficient in the Indian tongue.

His Lordship: I recognize that. I don't see, however, how the word for fire sled ties into anything. I think it gets a little removed. If he wanted to say, or if you wanted to ask him the questions about whether or not treaty money or anything like that, that is one thing, but to go through a whole series of words that seem to have no relation to what we are talking about, I don't know why we are doing it, that is all. I must say we are at Number 258 and I have yet to really understand why we have to go through 258 places. Maybe there is a point to it, and I think I can see a point or perhaps two points, but —

Mr. Clark: There is a whole bunch of them.

His Lordship: He seems to be going into such fantastic detail and I feel that it is something that might have been able to have been handled much more briefly. I have let you go on with this, as in many other things, hoping that I can understand exactly what you are getting at later.

Mr. Clark: It is sort of like, this isn't marginal; this is main line stuff. I don't mind explaining to the Court why it will fit together later.

His Lordship: Well all right.[7]

Lifeline II — Taking Flight

AS CRAIG'S STAR WAS RISING AT CAMP KANDALORE, IT TOOK A second run at Grade 13 at York Memorial Collegiate Institute for him to discover a way forward into the world of post-secondary schooling and toward a possible career track that would suit him. It may have been that his focus on woodsmanship skills and wilderness route finding, and his growing leadership skills and influence at camp, helped him to realize there was no shame in moving away from learning chemistry and the esoterics of French, Latin, and German and toward unpacking in school some of the mysteries of the natural world that had been revealed to him through his canoe travels. Craig's "victory lap" at York Memorial involved taking courses that would boost his grades but also, significantly, a course in biology with Mr. Pendlebury, which, it seems, set him on fire.

But it must have been hard for him to go the distance in the Toronto school system. This time he finished Grade 13 with flying colours and was offered a place in the biology program at the University of Guelph. But despite the relief he might have felt, the fact that his younger brother, Rod, who had skipped a grade in elementary school, was graduating at the same time must have a been a bit of a downer for Craig. Rod, like Craig, had the Macdonald brains, but he'd taken to formal schooling much more readily. He also had a seriously competitive streak that, however subtle — or not — it was, would have been difficult for Craig to ignore or avoid.

Craig's sister, Sandra, has a vivid memory of an event that happened on a family ski trip to Limberlost ski area near Huntsville, around the time her brothers were graduating from high school, that let the world know of the many and occasionally rancorous dimensions of her brothers' love for each other.

Hanging on a Limberlost resort cabin where they were staying was the biggest icicle any of them had ever seen. Craig, being Craig, decided this fantastic thing was so cool, he really needed to take it home to Toronto for further investigation. So he carefully removed it from the eaves of the cabin and was about to wrap it in a blanket for transport when Rod purposely and maliciously smashed the icicle. Sandra told me,

> They got into the most vicious fight I'd ever seen. I don't know if I'd ever seen them actually physically fighting before, but there were split lips, bloody noses, you know, hair torn out. The whole thing. This was a really serious fight which scared the hell out of me. My father stood back. There were people at the resort watching. And everybody was saying or thinking, "What do we do? What do we do?" I'm sure if there were police around, the police would have intervened. Eventually our father got them apart and we quickly left. Normally, I would have sat in the front seat of the station wagon because I was prone to carsickness, but on this occasion my mom and I sat in the back with Craig, and Rod sat in the front with our father.[1]

As she told me this story, Sandra added, "This moment stands out. It was horrible because I was so terrified by the fight. And, before talking to you, I asked Craig about it. Would he be okay with me telling you this story? 'Is there anything you don't want me to talk about with James Raffan, like the icicle story?' Craig replied: 'What's that?'"

Still, they were brothers with different interests and on separate agendas. But it must have been difficult, perhaps even a touch humiliating, when their two-year grade separation vanished and the two of them graduated from high school together in June of 1966. It was probably not an accident that Craig went straight to camp after finishing his exams and did not attend his high school convocation in the fall. The silver lining of this milestone, and maybe evidence of a measure of relief for his parents, was that Craig's mom gave the boys a sixteen-foot canvas-covered red Chestnut Deer canoe for their graduation. Craig was so happy with this graduation canoe that he personalized it by burning an Indigenous design he'd found of a snake eating a turtle egg on the bow.

In the fall of 1966, while Rod opted to go to York University to begin what became a distinguished career as a legal scholar,[2] Craig went the other way, to the University of Guelph, where he enrolled in Wellington College to study biology. Although technically the Johnston Hall dormitory at U of G was his first address away from home, he had spent enough time at camp by then for this to be an easy and welcome move toward increased independence for him.

With his broad knowledge of natural history, now buttressed by passions that Mr. Pendlebury had kindled in him for the formal study of biology, for the first time in his academic career, Craig thrived in the school setting. Guelph was originally an agricultural college and had more of a rural vibe than, say, the University of Toronto, McGill, or any of the bigger, more established universities to which Craig could have applied. So in addition to courses in physics and chemistry, Craig was able to learn about soils and plant pathology. He even got involved as a keen undergraduate in a hagfish experiment, having to do with fish behaviour, algae, and water surface tension. He even, for a time, joined the rifle club.

An extracurricular highlight of his four years at Guelph was meeting, through Kirk Wipper, Alex (Sass) Peepre, a running coach with the U of G Department of Human Kinetics. When Craig first went to Kandalore as a camper in 1959, the Intermediate Section head was John Swinden, who, with Kirk's urging, introduced orienteering to the camp. Craig loved to run, particularly cross-country, but when he decided he didn't have the lung capacity to be a champion, he discovered, through John Swinden, running in the bush with a map and compass. Kirk, who was a professor of Physical Education at the University of Toronto in addition to his role as owner and director of Kandalore, knew Sass Peepre as a colleague and kindred spirit, and knew that Craig would enjoy meeting him and perhaps continue his interest in orienteering at Guelph.

Craig did get interested, but it was a short-lived affair. Here's how Craig described the beginning and ending of his competitive orienteering career:

> Sass advertised that he was going to host the first competition for Canada. It was held at Corwhin [a little ski area just south of Guelph]. He had us in the hills back there, running all through the woods. He put a lot of work into setting up a course. And he encouraged me to get a team together. I got my brother out of York University. I got Brian Law, another camp friend, and another fellow U of G student, who went on to work for the World Health Organization in Geneva. We formed a team and actually had the winning score for Ontario teams, but there was a Quebec team that beat us by a few seconds. There were all sorts of teams there. Teams from the U.S. Marines at Fort Drum in New York State. They came up to do this competition as well. It was quite an event. This was my first and last orienteering event.

Craig was not much for team sports, or organized competition generally. But orienteering requires a very particular and well-honed set of map interpretation skills, and although the competitive aspect of what he learned from Sass Peepre fell by the wayside, Craig's eye for a contour and other topographic features only got sharper as a result of these experiences.

With the university year ending in April, Craig had a two-month gap between school and camp in July and August. He answered an ad for a seasonal naturalist's position at the Metro Toronto and Region Conservation Authority (MTRCA). Taking one look at Craig's impressive curricular and extracurricular pedigree in biology, woodsmanship, and wilderness travel, they gave him an interview, where they got a sense of his abilities as a natural teacher and raconteur. They hired him on the spot, even though they knew full well that he would be heading off to his staff role at Camp Kandalore in late June.

And so began Craig's lifelong connection to the world of outdoor education. It was a perfect fit. Throughout his university years, he served as a naturalist at conservation areas including Boyd, Kortright, Greenwood, Glen Haffy, and Albion Hills, teaching school kids mostly but doing public presentations as well, no doubt on technical and interpretive topics he had honed in his work at camp. He was not fussy on the uniform, but he loved the teaching. And through this work Craig became acquainted with many people amongst the MTRCA staff. Those who were influenced by Craig and his passions in these early days of his career, particularly teachers who brought their classes to the conservation areas for field trips, created a life-long professional network that's still humming for him well into retirement.

The rhythm of his university studies was punctuated each summer by a stint as a park naturalist for MTRCA that rolled into his guiding and canoe route research for Kirk Wipper at Kandalore. His life now was different than his experience growing up at 20 Westacres because what he did for fun and what he had to do for school were more or less completely aligned. That he was finding his stride as a student, naturalist, and biologist was noticed by his professors at Guelph, one of whom suggested that he consider continuing on with his own research in the context of a graduate program.

And that was all the impulse it took to see the rocket man from Westacres Drive in the Borough of York accept a full scholarship to move to California, perhaps inspired by his dad's and Uncle Keith's stories about driving the Pierce-Arrow out there during the Depression. He graduated from U of G, making the Dean of Biological Science's honour roll, and embarked on a two-year master's program in fisheries research at California State University, Long Beach, working out of the U.S. Naval Weapons Station at Seal Beach.

Graduate research was more demanding and time intensive than his undergraduate work, so he had to back off on the MTRCA contracts while he was based in California. But Craig still managed to return to camp each summer to continue the work that in some substantial and significant ways completed him. In 1972, Craig completed his master's research and produced a thesis entitled *Aspects of the Life History of the Arrow Goby,* Clevelandia ios *(Jordan and Gilbert), in Anaheim Bay, California, with Comments on the Cephalic-Lateralis System in the Fish Family Gobiidae.*

Whatever this stint chasing goldfish in the sun and surf of Anaheim Bay did for him intellectually, personally, or socially — he did actually learn to surf, he said — he was well paid for the work and he garnered respect for the research he conducted as part of a team working within a larger conceptual framework. But while he was in California, there was always something — or some*one* — he'd left behind in Guelph that was very much on his mind.

Although men and women stayed in separate residences at U of G in those days, everyone ate their meals in the cafeteria in Creelman Hall, and it was here that a winsome first-year brunette named Doris Henderson caught Craig's eye. Sporting a Beatles haircut, in his jeans and checked flannel shirt, he checked her out for a few days before he eventually mustered the courage to sidle up to her one day and sit down with his lunch.

In that first face-to-face encounter, Doris — who was a year older than most of the other first-year women because she'd been to teachers' college before starting a general arts degree at Guelph — told him that she'd been home to the farm near Palmerston that weekend and that she was a horse lover. But this past Saturday, she'd had the chance to ride a donkey, she said.

Craig, being Craig, twigged to this topic immediately and, from somewhere deep in his overflowing mental filing cabinets of biological minutiae, he launched into a detailed unpacking of the difference between a "hinny" and a "jenny" — the former being a hybrid cross between a male horse and a female donkey, and the latter being a female donkey. Horses, he explained, have sixty-four chromosomes while donkeys have only sixty-two, leaving hinnies, or mules, with sixty-three, which usually prevents the chromosomes in a zygote produced in a horse and donkey cross from pairing up properly and creating successful embryos, which is why many of them are sterile. If you please.

In May of 1969, as Craig went off to his naturalist's job with the MTRCA, Doris was offered a maternity-leave teaching position in Harriston, which turned into a full-time job the following year. She didn't go back to university to complete her degree until much later in life. But the two of them dated during Craig's fourth year and, by the time he'd accepted the scholarship to Cal State Long Beach, their relationship was well established.

Doris described that separation:

> We wrote back and forth. Craig got rid of my letters but I still have his. We phoned maybe a couple of times

> during the year. And then, of course, he came home in the summer and went to camp. So he was gone because he was guiding those big trips. But we were serious. My cousin, who I was very, very close to, said, "Doris, don't wait on him." But I told my cousin that I was going to give Craig two years and see what happens. There was another teacher that came to the school where I was working and we could have done a relationship. But I just balked and said I'll wait the two years. And, yeah, Craig was faithful too.[3]

Although Doris's side of the letter-writing romance that blossomed while Craig was away is lost forever, what happened inside that epistolic courtship is revealed in nearly fifty letters that Craig wrote to her between May 1970 and August 1972 — letters kept by Doris for posterity in a small black stationery box. They wrote biweekly. In this little treasure trove of insight into their courtship is one letter written a few months before Craig left for California.

In February 1970, as Craig's final year at U of G was winding down and Doris was teaching in Harriston, Craig was admitted to St. Joseph's Hospital in Guelph with persistent infections in his gums and teeth that needed intensive medical care. In pencil, on a utilitarian single-fold kraft paper towel presumably taken from a hospital washroom, Craig showed a side of himself that to this day, I think, only Doris has experienced. Although much more restrained than later letters, it was clear that Craig was smitten from the very beginning with the young teacher from the farm near Palmerston:

> Dear Doris:
> If I could return even a fraction of the happiness you bring to me, it would be certainly worthwhile.
>
> You are more precious to me than all the treasures of the world. I yearn to be with you. In this "prison," the minutes have turned to hours and the hours into days. You are constantly on my mind.

Eventually the problems that presently surround us, will be resolved. I dream of the time when we will be truly together and able to indulge in our mutual pleasure.

Doris I need your love and miss you greatly.

Craig

Craig went off to camp that summer, and although he did send an invitation by mail to Doris to attend his U of G convocation, the correspondence doesn't really begin until he departed for California that fall. They tried talking over the phone, but long-distance international calling was very expensive at the time. Also, Craig lived in a student residence in Long Beach that had only a pay phone at the end of the hall, audible to prying ears and possibly resulting in taunts for long-distance Romeos gushing over the wires. On October 10, 1970, he wrote about this awkward situation: "It was wonderful talking to you over the telephone. It didn't seem like two thousand miles away. Since I was using a wall payphone and people were standing around I could not be very personal and tell you how much I really love you. California is a nice place to visit or go to school but I would not care to live down here especially without you."

But in a letter written the following week, he responded to Doris talking about the interests of the "other teacher" who was circling her and exploring possibilities of a relationship. There might even have been a slight pang of jealousy in Craig's matter-of-fact analysis of that situation. He wrote: "In your letter you mentioned a teacher with a horse. My advice is to be honest with him. If you want to see his horse do so. Make it clear that you would like to be good friends with him, but that is as far as it goes. Tell him you are very close to your boy friend and you don't want to lead him into disappointment."

His letters to Doris from Long Beach detailed the courses and field trips he was taking, news of his roommate, general comments on what the "hippies" on the beach were doing with their surfboards, and general patter about the minutiae of his life as a graduate student in marine biology at a prestigious American school. As the 1970 fall term wound on, he

When doing his master's degree in California, Craig wrote faithfully every week to Doris and, on this occasion, he included some self-portraits taken with his roommate's black-and-white film camera.

realized his finances wouldn't allow him to come home for Christmas. It was clear through his references to repeated attempts to phone and by the tone of his letters that he was getting increasingly homesick for her.

On December 18, he apologized for calling at midnight instead of 11:00 p.m. because he'd gotten mixed up about the time difference. He admitted that this was the first time he set down his feelings directly, without doing a rough draft first. He wrote: "You will probably notice that my writing and ideas are a bit fragmented in this letter. Usually I make a rough copy to you. However, in this letter, I am trying to learn to write my thoughts down without making a rough copy like you are able to do." Then he wished her a merry Christmas, and signed the letter, "With loads and loads of love, Craig." Knowing that her birthday was coming up, he spelled out "Happy Birthday" on the bottom of the page in *x*'s and *o*'s with hand-drawn hearts liberally sprinkled through the text.

Craig's love of land and geography came through in descriptions of travelling with his classmates and research colleagues through southern California. He told Doris of horse trails winding up into the high chaparral of the Santa Ana Mountains. And, as the warm winter of 1971 progressed, he allowed as to how he was missing snow but thinking of the canoe trips he'd lead at Kandalore in the coming summer. Although he was not a guy to run others down in conversation — or to speak ill of anyone, really — on April 25, 1971, he confided to Doris that "Kirk Wipper, my summer boss, has cut me short $100.00 on my contract, so I will have to hassle this out with him. His 'game playing' makes me very angry."

What was new for Craig, in this phase of his life, was a relationship with the sea. In that same handwritten April missive, he described one of his regular sampling trips out into what he called "the deep waters of the Pacific":

> We were in a storm with 30 knot winds. Quite a few waves crashed right over the deck and nearly knocked some of the crew right out of the boat. It was so rough that you could not even walk on the deck. You had to crawl or hang onto the railings. Some of the waves were as tall as your house. Needless to say the skipper was quite scared and cut the trip short. Boy, was I happy to reach land. I don't think we will be going out again when it is so rough (it was a real rolly-coaster ride both up and down and sideways).

Although he spent almost all of the summer of 1971 leading trips for Kirk Wipper, Craig and Doris were able to get together a couple of times. And, in the correspondence that resumed on his return to California that fall, it was clear that their love for each other, now tested again face to face, was deepening and that their relationship was maturing. Although we don't know how Doris expressed her concerns about how her life was unfolding, on Saturday, September 25, 1971, Craig reached out to her in the most touching way, buoying her spirits and affirming his love:

> On Oct. 8th I will be on the better side of 25. Imagine that! ¼ of a century! I will try and phone you either Friday Oct 8th or Sat. Oct 9th somewhere between 10:00 and 10:30 P.M. your time. You tell me which day is best for you. Remember I love you *very* much and I am just waiting for the time when we will be back together again. I have all the confidence in the world about your abilities and someday soon you will realize that I am right. It just takes a little time to build up some self-confidence so you are able to try hard and do anything you desire — and do it well. You are the girl of my dreams — I believe in you. Just remember in whatever you are doing I will always be at the sidelines anxious to cheer you on because I know that you can do anything you wish.
>
> It is wonderful to know that somebody believes in what you are doing and cares for you. No man has ever loved a girl as much as I love you. It is so strong at times I can feel it through my whole body. I can describe the feeling. It is like a huge magnetic force stronger than steel yet at the same time warm and soft.

If there was a crescendo and turning point in this remarkably revealing correspondence, it was a letter written on November 16, 1971. It started out in the same way most of his letters did, with pleasantries of the day-to-day: "I was up at UCLA and you could see for miles out on the Pacific Ocean.... I've written my parents about the possibility of you coming along with them.... I hope your father is feeling better after the accident? Was the tree one of those old elms? They call them 'widow makers' were [*sic*] I come from," and so on. But then, out of the California blue, Craig unleashed a poetic unburdening of his soul that, although not actually including the phrase "Will you marry me?," got as close to a proposal as Craig ever came.

Doris and Craig were engaged soon after he finished his master's degree and returned to Ontario. When Doris spoke to me in preparation for this book, I knew nothing about the contents of this correspondence and I asked,

"Did he get down on one knee to propose?" Doris laughed heartily and said, "Heavens no. I just said, 'Craig, I need a ring.' There's no romance to Craig Macdonald. Zero romance to Craig Macdonald. Except that he's extremely faithful.... I looked at him and thought, 'Okay, I know he's going to be steady.' I just felt he had a purpose in life. He was an outdoorsman in one way, but I was an outdoorswoman in another way. I was very much a farm girl. I was very much a horse girl. He never asked me to give up my horses. Or anything like that. I think that's the way marriages should work."[4] As the correspondence so clearly and sweetly illustrates, this is a relationship that is much more about profound internal heart-to-heart bonding than about any kind of visible expression of affection.

Upon his return from California, Craig went back to the MTRCA, but it wasn't long before he accepted a contract from the Ontario Ministry of Natural Resources (OMNR) to help research Ontario canoe routes for a guidebook project. Commuting back and forth from his parents' place to the farm near Palmerston, Craig worked out of an OMNR office in Richmond Hill and then moved down to Queen's Park, where he immersed himself deeper than ever in the archival maps and canoe route research he had started for Kirk Wipper at Kandalore. Now charged with examining the watersheds and ancient travel routes for the entire province, he was particularly struck by a map hand-drawn by eighteenth-century surveyor Augustus Jones, which showed the position and original Indigenous place name of every creek and river running into Lake Ontario from Toronto to Niagara Falls.

Doris and Craig were married in a very small ceremony at the farm on August 11, 1973, coincidentally just four days after a far-reaching legal land caution, which would in many ways shape the arc of his life, was filed by the Teme-Augama Anishnabai against the Government of Ontario; this claim of ownership by the people of Bear Island would suspend development of their Temagami homeland until title was formally established by the courts. Doris gave up her teaching position in Harriston to support Craig in the start of what would be a full career with the OMNR, and the newlyweds moved into an apartment on Glenlake Avenue near High Park.

Craig loved the work at Queen's Park but hated the head office environment. His first task was to write the proceedings of the first Ontario Trails

Conference, a multi-volume effort. This was followed by the launching of the Ontario snowmobile program. This included drafting and enabling legislation, developing provincial trail standards and signage, coordinating trail development and grooming operations, provincial acquisition, and dispersal of grooming equipment, and issuing over one million dollars in trail grants to snowmobile clubs. Nasty office politics evolved between Craig and his superior at Queen's Park. While all this was going on, Doris, who did a bit of volunteer teaching at Enoch Turner Schoolhouse, was bored, lonely, and decidedly out of place in midtown Toronto, missing the farm, missing her family.

As a solution, Craig was relocated to the OMNR Regional Office in Huntsville on a new two-year contract as a "recreation specialist" (a job title he would carry right through to his retirement from the OMNR forty-two years later), hired to help figure out how to set up a provincial trail system. But, being out of Toronto, he could continue his trail research and other related work. He rented an office in Huntsville and used this opportunity to compile his field notes and to advance his work on a preliminary hand-drawn version of the Temagami historical map that would assist him with the interview process in the future.

The move solved the Toronto problems for Doris. She'd never really been north of Wellington County, except on holiday, but she was a can-do kind of person who felt a palpable sense of emancipation in that move from Glenlake Avenue to a rental place at 22 Cliff Avenue in Huntsville. She has a vivid memory of the drive north, of getting up past Gravenhurst on Ontario Highway 11 and smelling the pines and exclaiming to Craig, "Oh my gosh, this is beautiful!"

By now, the sheaf of charts he'd rolled in the varnished cardboard travelling case for the 1967 Lake Superior journey had grown, thanks to the canoe route research he'd been doing for Kirk and then for the Ontario canoe routes book that was being published by the OMNR. It was now a space-eating welter of tubes, rolls, stacks, and envelopes. There were boxes full of his maps and handwritten notes from interviews with people like Pete Albany at Mowat's Landing and a growing list of informants at Bear Island, Golden Lake, and other First Nation Reserves from Quebec to Manitoba through mid-north Ontario. In those early days in Huntsville, Doris was pregnant with their first child, and the only way for Craig to spread out his work — which was really the beginning of the creation of the *Historical Map of Temagami* — was to rent a bigger place in downtown Huntsville.

Craig was hired by the OMNR to figure out how to set up and run a provincial trails system. The idea had been launched to inventory all the existing trails; to liaise with the various user groups, like snowmobilers, cyclists, cross-country skiers, hikers, ATV enthusiasts, and naturalists; and to talk to the huge mix of landowners, including private citizens, municipalities, and First Nation communities. George Moroz, his boss in Huntsville, described the young biologist who came up from Toronto this way: "Anything he took an interest in, he was obsessive about it. He'd take it on and there wasn't any detail that he didn't pursue, to make it as perfect as he thought it should be. He was thorough in everything that he did. Meticulous. In some respects that would drive some people mad, to have to deal with him. No stone was left unturned."[5]

By the time this contract with OMNR ran its course, with no possibilities for extension or rehire at that point, Craig didn't miss a beat with his research. He had the bit in his teeth and was running with it. Prompted by his early conversations with Pete Albany, Charlie Haultain, and some of the old-timers on Lady Evelyn Lake, Craig had experienced an epiphany, particularly with respect to trail and travel knowledge held by Indigenous Peoples he'd encountered:

> I realized the frailty of human information. The Native people were basically an oral tradition. They hadn't been writing stuff down because they didn't have a written language. So, it wasn't in their mindset to do that. It's just over time I'd come to the realization that this trail and travel route information I was gathering was important. Back in 1968–69, I thought, "Whoa! There's something here." And then I went at it. Once I started doing the interviewing, I became convinced that there's something really here that was very worth pursuing.[6]

This realization was further informed by the work of Augustus Jones and by his own forebears, Sandford Fleming, John Fleming (who was on the Hind Expedition), and his grandmother's father, who was Head Commissar for the Nipigon to White River portion of the CPR. No doubt he was also inspired by other cartographers, ethnologists, geographers, and travellers he had encountered in his research.

From the moment Craig launched into this lifetime quest, particularly when he started travelling hither and yon to talk to trappers and Elders, Doris backed him as a willing and enabling partner by giving him the freedom to follow his bliss. Their daughter Nancy arrived on September 18, 1976, and Doris synched her days with the rhythms of the baby, making it possible for Craig to continue with his growing obsession. A very capable mother and housekeeper — and a qualified and experienced school teacher — she set the pattern of their life in their little place in Huntsville by putting her teaching career on hold and taking on the essential roles and responsibilities of parenting, keeping communications with the Henderson and Macdonald grandparents going, shopping, cooking, cleaning, and socializing with the neighbours, while Craig brought in a regular paycheque and pitched in when she asked him. Mostly, however, his time was unfettered with any domestic chores, which left him free to focus on his work and his research.

Craig had first encountered Pete Albany in his wall tents at Mowat's Landing in the 1960s, when he was researching canoe routes for Kirk Wipper. In June 1977, Craig found his way back to Peter's place for a more formal interview, which would provide him not only with a treasure trove of place names inside n'Daki Menan but also with names of people at Bear Island who could add to Craig's knowledge base. As far as permission to access Elders on Bear Island was concerned, by this time Craig had also met Chief Gary Potts, who, having filed the land caution on behalf of the band five years before, was very keen on Craig continuing his work, knowing it was essentially creating an evidentiary base for the land claim trial that would eventually come.

To get an inside look at Craig's interests and his method of interviewing and recording information, here is a sample of his notes, which were written

Peter and Josephine Albany had a campsite adjacent to Mowat's Landing on the Montreal River in Temagami, where Camp Kandalore canoe trips in the area began. Peter was Craig's first real informant in a decades-long investigation into Indigenous winter and summer travel and travel routes.

in pencil on foolscap (probably in the car) immediately following his two-hour conversation with Pete Albany. These notes illustrate Craig's singular focus on place names and routes, particularly the winter trails, which are different from the summer trails. But, given that the notes were written after the interview, with just a few pencil scratches on a topographic map made during the conversation, they also point to Craig's remarkable memory for geographic detail.

> Peter Albany — Mowat's Landing
> Interview Sat. June 18, 1977
> – Peter Albany's father came down with Missionary from the Albany River, he stayed down here, never returning (his name was also Peter)
> – Peter had two hand sleighs for drawing wood (one broken and no longer used with 4 inch wide runners made from yellow birch coming from the Willow Island logging camp.
> – the new sleigh is 16 inches wide and quite long possibly 5 ft., bottoms of runners covered — raves and runners — ash, legs — elm — makes his own snowshoes out of ash — very similar to Clarence Bogue's largest design of flat-toed polywogs.
> – family one son, the rest daughters one living in Hearst, another in N. Bay. His sister Jane Katt of Bear Island
> – worked for 18 years for lumber co. — at night in winter on icing the roads, during day in summer captaining a gas driven (gaiter) on Lady Evelyn L.
> – claims that the high concrete dam in 1925 which replaced the old wooden dam — sent Wendabin's Cabin as well as the Ranger cabin floating. — the Ranger's cabin was located on a long island just south of Obisaga near the east shore.

Negigo Z.   Otter Ck.   Barber Creek

Kapkigdownkog ________? Big Spring Lake
Spring (fountain) mokid ji wanibigBaraga

Confirmed the bibon. Milean route from the Montreal R. through Big Spring to Barber Ck.

Kokgiakajin ________? Indian L.
Tom Moore lived on the N side in the narrows of this lake before moving down with Charlie Mowatt when he got too old to stay alone.

Ka-kino je-kog Pike L.
Small unnamed lake south of Lepha L. on bibon, kana route from Mowatts to Indian L.
Bad ice opposite cabin located on East shore of Montreal R. just down river from the outlet of Indian L. ← the bibon.mikan was not only shorter but avoided this bad ice.

Akik-kanab-agoding ________? Schumann L.
Akik — pot kanab — hiddenagoding — hanging
refers to a pot that was hung up as a portage marker

Wawiaygama Pawatik round L. rapids
Mountain Chutes

Kaginoganok
Largest lake on Lady Evelyn bibonkana in middle KLOCK TWP. (Large island in eastern end)

And so on for eight more foolscap pages.

When Craig's two-year contract in Huntsville expired, the provincial trail system was under way but the research imperative was all-consuming. Doris was working full-time looking after the baby. She'd learned on the farm how to stretch a buck and to make ends meet. But they needed an income. Craig did get a modest Wintario grant to pursue his map project, but grant money was not really meant for rent or groceries. He was travelling all over, interviewing people, but he had to get another job. Soon. He was still very much connected to the informal network of OMNR staff and contractors, and a staff job came up at the Leslie M. Frost Centre, an old forest ranger training school near Dorset, Ontario, that had been repurposed as the province's first outdoor education centre dedicated to environmental and resource management.

Craig's resumé at the time was very research and policy oriented, detailing his formal education and the work he'd done under contract with the ministry, but this recreational specialist position at the Frost Centre was much more like his camp positions. It called for the kind of teaching and front-end interaction with the public that he had done for the MTRCA. Doris took a look at the dossier he was proposing to send and said, "Craig, I don't think this is what they're looking for. They're looking for somebody who can teach."

That had not occurred to him. With Doris's domestic pedagogy and continued assistance as chief adviser and in-house editor, Craig rewrote his resumé, highlighting all the work he'd done for Kirk at Kandalore, working with kids, and his teaching at the conservation areas around Toronto. They travelled together to the Frost Centre. While Craig sat for an interview, Doris toured the site and met the families of some of the other staff members. And *voilà*! They got the job. And with that, in the fall of 1977, the young Macdonald family moved into a modest staff house on the Frost Centre Campus on St. Nora Lake.

The job of recreational specialist at the Frost Centre had many dimensions, from teaching and trail maintenance and track-setting for their cross-country ski program to higher order concerns like program development, provincial land use standards, and coordination of communications and information flow amongst the various arms of the OMNR, including conservation and parks. But in many respects the job was perfect for Craig because he was busy, outdoors, and working with his hands a lot of the time. A good portion of the work he took on related to managing recreational use of the 24,000-hectare property and to his continuing avocational interest in trails and travel technology, particularly traditional snowshoes, sleds, toboggans, and winter camping techniques. His boss, George Hamilton, told me: "You never knew what Craig was going to be doing, but you could always count on him to get it done. He was a pretty unique guy. He lived and breathed camping. Always fanatical. And people liked listening to Craig. He was popular with the students who came to the Frost Centre."

Through his conversations about traditional trails and travel techniques with people like Elder Donald McKenzie at Bear Island, Craig had learned a great deal about the technology associated with canoes, maple sugaring, trapping, and traditional life. He found himself particularly drawn to the mechanics of how Indigenous people travelled in winter. They described to him special winter-only trails, called *bonkanah*, which he duly marked on his maps, but there was also a vast body of knowledge built into the snowshoes, sleds, toboggans, tents, tumps and harnesses, and other homemade paraphernalia used to travel and live comfortably in the winter woods. Craig stored his growing collection of these items in the basement of their little house at the Frost Centre, but in his work he also used the Frost Centre's considerable resources, including access to wood and commercial shop facilities, to start improvising and practising some of the building and fabrication techniques he was learning about in his interviews.

In their second winter at the Frost Centre, Craig hosted a winter camping workshop that brought together three dozen or so educators and outdoor enthusiasts to share their technologies and techniques for being comfortable below zero. One of the organizers was Jim Wood, outdoor education specialist from the Muskoka Board of Education, whom Craig had met at Albion Hills Conservation Area when Craig was a summer naturalist for the MTRCA and Jim was a staff teacher at that outdoor education centre. The co-lead on this workshop was an ambitious Huntsville teacher, Roger Bragg, who had been dabbling in winter camping with his students and was totally fascinated with Craig's research and what he was learning about trails, maps, and winter travel technologies.

Craig and Doris's next-door neighbours were Joe Stocking and his family. Joe had been part of the first class of forest rangers trained at the Frost Centre in 1947. Having grown up in the bush north of Red Lake, in far northwestern Ontario, Joe had a lot of woodsmanship skills and trail savvy even before he began his hitch as a towerman and ranger for Ontario Lands and Forests. But now, in semi-retirement, he was part of the über-knowledgeable staff at the Frost Centre. Craig, as he did, took it upon himself to formally interview Joe about his knowledge of trails and camping techniques, particularly for winter, including using Egyptian cotton canvas wall tents heated with portable tin stoves. As Craig filled in the gaps in the things he was learning from his informants by improvising (in reality, he was making things up), it was his neighbour who helped him understand that the little commercial tin trash burner that he was using to heat his tent might be replaced with a purpose-built rectangular stove for which Joe had the plans in his head.

Following lines of inquiry from his research, Craig also found his way to Jack Saran, the proprietor of Northwest Canvas in North Bay, who agreed to sew up a nine-by-twelve-foot white canvas wall tent with a special burlap "sod cloth" sewn along the bottom of the wall. The sod cloth helped make a seal with the snow and would keep fuel consumption down on the coldest nights of the winter. As the date for the winter camping workshop organized by Roger Bragg and Jim Wood was approaching, Craig had his new tent set up on the lawn outside his house at the Frost Centre. Anybody who came by was invited in for a hot cup of tea or to try a sleepout to see what this "hot" camping was all about.

Just before people were to gather at the Frost Centre for the January weekend, Roger got sick, so Craig was left to fill the Friday night program slot with a talk about snowshoes. There were historical snowshoes in the Frost Centre, but Craig also trotted out his own growing collection, gathered on his research trips and travels. This may have been the first of a lifetime of presentations on snowshoes that Craig is still doing to this day, and that he'll be doing into the future as long as he is able and people are asking.

But this was not an indoor-style set of lectures. It was a workshop, a chance to get out there and to share. So the following morning, everyone packed up their kits; loaded their makeshift sleds, pulks, and toboggans; and strapped on their snowshoes. They streamed down to the ice of St. Nora Lake, where they fanned out in groups to build a number of overnight camps, where they'd all spend Saturday night. In conversation with me, Craig described what happened:

> One group made a bivouac by the falls on Sherborne Creek. And there was another group that went to Sundew Portage. Jim Wood had a one-man Hap Wilson–style thing where you just tied a pole to a tree, leaned it down, took a tarp and then put it over the pole and then just stuck a wood stove in [beside it]. And then they used one of those Canadian Tire space blankets or some darn thing to close it in.
>
> Another group, maybe eight or nine of them, made a lean-to structure up against a cliff. Roger had recovered by Saturday noon and he went out with his Calvin Rutstrum wedge tent. Another guy, who came from the Peterborough/Lindsay area, brought his own wall tent, like mine. He housed a bunch of people. And then I went down to Otter Pond down toward Silver Doe and Silver Buck Lakes and housed another five or six in my nine-by-twelve-foot Northwest canvas tent. And once all the camps were built, we had a get-together and toured the structures.

Because Craig is so incredibly understated about the length and breadth of his achievement, it may well be that the attendees of this first winter camping workshop at the Frost Centre did not really comprehend the authenticity and veracity of the technology and techniques Craig taught them over the weekend. By now, through his work travels, he had interviewed dozens of Indigenous and non-Indigenous knowledge keepers. He had carefully noted their routes and place names on his maps. In the name of accuracy, he had challenged doubtful things that people had mentioned or topics about which his informants might not have been totally clear and had cross-referenced them with multiple informants. And, with the help of Joe Stocking and others at the Frost Centre, he had attempted to create or re-create their all-wooden toboggans, stanchion sleighs, and snowshoes.

As Craig had known since his days with soapbox racers, kites, and rockets, the only way to reveal the true learnings from any technology was to try it out; to experiment with dimensions, designs, and techniques; and to keep notes on his findings. So, from the Frost Centre's front lawn, Craig started going on winter journeys through the trails that were part of his responsibility as recreational specialist. He did solo bivouac journeys and, through his circle of friends and cronies, including members of the Wilderness Canoe Association, he started taking people along with him to try out different configurations.

For snowshoe bindings he started with commercially made leather Bates human harnesses and ordinary hard-rubber-bottomed winter boots. But he soon realized the motion and fluidity required for efficient travel on snowshoes. Based on his continuing conversations with old-time winter travellers, he eventually moved back in time to leather moccasins and cotton lampwick bindings. Soft footwear was not only warmer, but, when it was holding smooth lampwick bindings, a little flick and twist of the ankle would separate the feet from the snowshoes in the event of an accidental dunk in cold water. With hard boots and leather bindings, Craig learned from first-hand experience that if he fell through the ice, he'd either have to put his hands (and possibly his head) into the icy water to unbuckle his snowshoes or he'd have to learn to swim while wearing the things.

A similar improvisational evolution occurred in the sliders that Craig employed to move his winter camping kit from place to place. Through his interview connections to the Algonquins of Pikwakanagan, and perhaps because Avery and Sons had been the main suppliers of snowshoes to Ontario Lands and Forests for years through their shop in Whitney, Craig had gotten a five-foot wooden sled from Alvin Avery. Needing more capacity than this beautiful little artisanal slider could offer, Craig took himself back to Canadian Tire in Huntsville for a longer manufactured wooden toboggan. Back in the shop at the Frost Centre, he narrowed the toboggan, so it would follow better in a snowshoe track, by removing a wooden slat from either side and shortening the crossbars accordingly. He tried wax on the bottom to make it slide more smoothly on snow, but that was not as successful as he'd hoped. So he attached pieces of coloured "crazy carpet," which children used for sliding in the snow, to the toboggan's underside, which worked quite well.

Soon he was back out in the woods where, following directions from his Anishinaabe informants, he harvested a white birch tree trunk, which he aged in a snowdrift before bringing it into the shop for cutting, planing, steaming, and bending to make his own wooden toboggans. Same with a succession of meticulously made wooden leg-sleighs, which took longer to fabricate because there was more design and more complicated joinery involved. In time, however, he realized through experience the fragility of these handmade reproductions and the prodigious amount of time it took to make them, and he began fashioning reproductions of Indigenous toboggans out of the polyethylene material used to shoe the skis of military transport planes. For his reproduction sleds he used plywood and dimensional spruce lumber with polyethylene strips on the runners.

Of course, what use would these sliders be if they did not take the maker out onto the *bonkanah* to learn about the character of these winter trails? To prepare for these journeys, he started planning in earnest. Craig reviewed his growing sheaf of pencil-on-foolscap notes from his interviews. And he scoured the books in his ever-expanding library for accounts of traditional winter travel. One of his particular heroes was Paul Provencher, whose first book, *I Live in the Woods: A Book of Personal Recollections and Woodland Lore* (1953) he was introduced to in 1960 by his dad, who was a personal friend of the translator of Provencher's second book, *Provencher: Last of the Coureurs de Bois* (1976). These volumes were near-biblical references for Craig.

To bring these references and information about tents and stoves to the attention of people interested in joining him on his research journeys, Craig was a faithful mail and phone correspondent for anyone with questions. But to save some time, in 1977 he drew up a multi-page mimeographed document called "Instructions for the *Keejab* Trail Stove." This included hand-drawn diagrams and typewritten text on seven topics Craig felt were key to understanding what hot camping was all about: erecting wall tents suitable for stove installation, stove set-up for wall tents in deep snow, stove set-up for wall tents in shallow snow, stove set-up for cabins, pre-operating directions for the stove, seven golden rules of stove operation, and cooking and baking on the stove.

With all the growing interest in hot camping and winter travel, in the fall of 1979 Craig reached out to the Wilderness Canoe Association to invite interested folks to join him in his most ambitious expedition to date, an east-to-west traverse of Algonquin Park from Whitney to Dorset on a succession of traditional trails. In the winter of 1980, there were eight of them in two tents, and for five days they slogged their way through pain of deep cold, and the joys and hazards of sharing a physically and mentally challenging journey with people of varying ages, tolerances, and abilities. But through this journey, Craig was able to show that Algonquin Park, which was empty of recreational users most winters, was a place where people could play *and* learn about traditional routes and travel techniques.

Although Craig did all the logistics planning, correspondence, and equipment preparation associated with this trip, Doris was totally in charge of planning and preparing the calorie-rich menu that would power the expedition members through the long days on the trail: packets of bacon and oatmeal for breakfasts; cheeses, breads, and jams for lunches; and stews, chilis, and other one-pot dinners. Their second daughter, Janet, had been born in July 1979, so Doris went about shopping, cooking, drying, freezing, and packing all of this food while feeding and entertaining a three-year-old and a six-month old. Her friends and neighbours from the Frost Centre, who were like an extended family, pitched in to help with child care. Doris's contribution to getting these research expeditions out the door and onto the trail can't be underestimated, even if she didn't always get the recognition she so richly deserved.

A scant 380 days after Janet was born, Craig and Doris's third child, Colin, arrived, making the Macdonald household just that much busier and chaotic. Even with a three-year-old, a one-year-old, and a newborn, Doris was so organized that she still was able to support her husband in his extracurricular projects. By now, Craig had all his maps, research materials, and other papers, as well as snowshoes, tents, stoves, axes, saws, canoes, and other equipment, neatly shelved and stowed in the basement of their little house at the Frost Centre. They were partners in child rearing, particularly when Doris asked for help, but with Craig's work at the Frost Centre and his day and overnight trips to do interviews in North Bay, Sudbury, Temagami, or beyond, sometimes it was easier for Doris to just let him go about his business with no expectation of steady help around the house.

Craig's work on trails at the Frost Centre kept him busy during the day, but in his head he focused on how to collate, gather, and create a cartographic picture of all the place names and knowledge he was now holding as a result of nearly ten years of intensive learning, a map that would honour and document the knowledge that had been given to him through the patience and kindness of his informants. When speaking with an Elder from a particular area, Craig would pull out one or another of his provincial forestry maps, which were bigger in scale than the 1:50,000 or 1:126,720 charts from the National Topographic System of maps. Or, depending on where he was, as he listened to an informant during an interview, he made a note on a larger-scale map of the general location and then would work from a hand-drawn sketch to capture a winter or summer trail. But even by the time he and Doris moved to the Frost Centre, Craig was clear that what he really needed to do was make his own map at a scale that would accommodate the granular detail of the trails, place names, water levels, and connections between watersheds. Craig mentioned his project to his contacts in the mapping branch of the OMNR, and it was through them that he determined that the largest-size paper that could be printed on presses and paper-handling equipment available to the ministry was about three feet by four feet. Craig decided that his finished map would be this size.

It was then just a matter of deciding how much territory to represent, which was all about determining what scale his map would be. He settled on a scale of 1:126,720, or two statute miles to the inch. This was an odd scale, to be sure, but based on the largest printing press in the country that was at Ashton Potter in Toronto and his desire to get as much detail onto

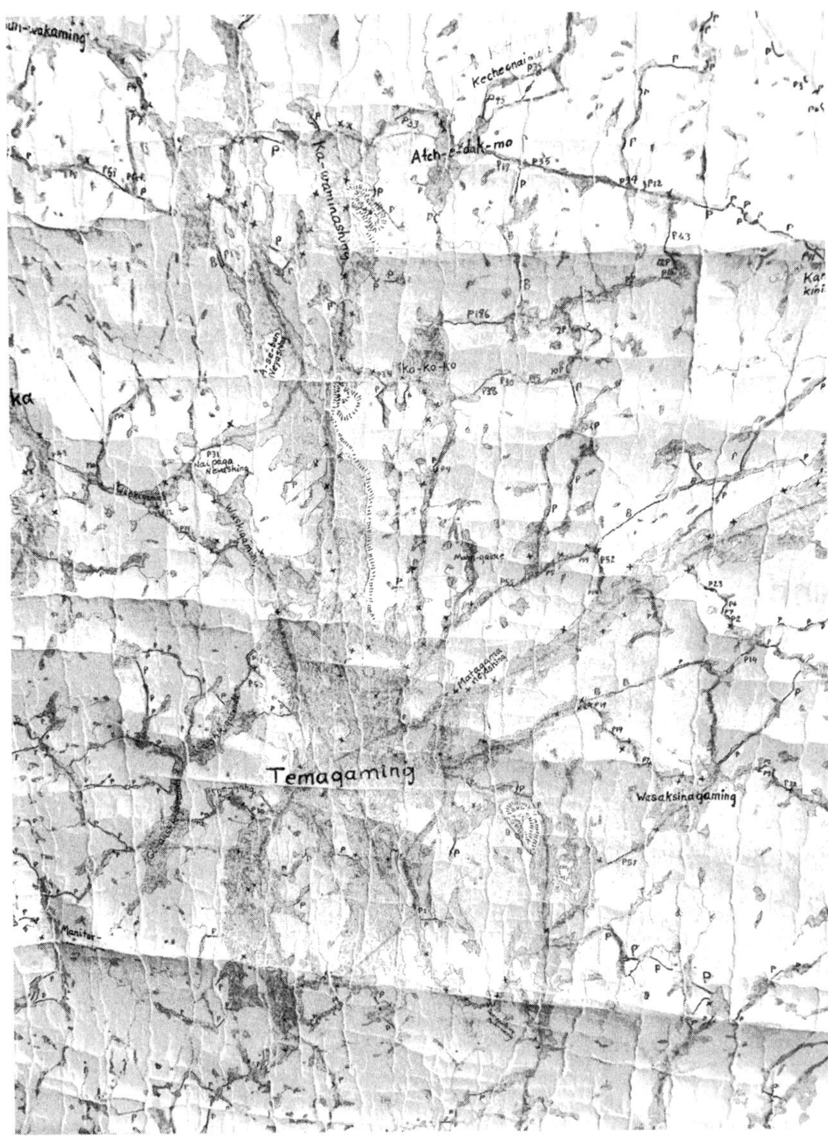

If there is a document that is more precious than the printed version of the *Historical Map of Temagami*, then it is the worn and rolled hand-marked topographic map that Craig took with him to interviews with Elders. Any informant who might have been slow to trust Craig would often be drawn in by the look and feel of the map and the information in the head of their interviewer.

the map as possible, it made sense. With big Lake Temeeaygaming centred on the page, this scale allowed ground coverage that reached from Lake Temiskaming in the east to Lake Wanapitei and the Keewaydino Zibi in the west, and from Tomiko River in the south to Elk Lake in the north. By now he had detail that went far beyond this area, but this covered all but the northern extent of n'Daki Menan, the traditional territory of the Teme-Augama Anishnabai, and this parcel of land was where the main concentration of geographic detail from his research was situated.

The intensity of Craig's growing passion for his research and for the map was not without its frustrations for Doris. "I would get really discouraged," she said.

> I would say, "Come on, Craig.... Do you have to go down and do the map right now?" He would come home, flake out on the bed, eat dinner, and then head to the basement to work on his map. That was every night. He never put a child to bed or anything like that. He never read the kids a story in his life. He did sometimes tell them stories.... But he never interfered in how I raised the kids. Now, maybe he thought I was doing an okay job. I made the decisions. And you know what? I was raised that way. My mom made all the decisions. The only time my dad ever interfered was — if he ever got involved — well, I knew I was really, really in trouble. Because I was raised that way, I felt that's the way we should run our relationship. I felt Craig's out making the money and my job is to run the house.[7]

Doris would confide in her mom from time to time. Doris told me,

> She was a very sensible woman. I'd complain to her and she'd say, "Well, Doris, you know he's not running after another woman. You know you're not dealing with force. You know you're not running to the bar to find him. You

know exactly where he is. He's in the basement." And Mom would say, "You've got to live with what you've got." And what would really help is every so often, when Craig would be invited to speak about his map, we would get a babysitter in and I would go with him. And then I'd hear about what he was doing in the basement. He would talk about his research. The people all raved about it, how wonderful it was. And then, I would come back and say, "Okay, let him go with this. I can run this household."

Travelling with Craig II — James Bay Bound

IT WAS IN CRAIG'S LAIR IN THE BASEMENT OF THE MACDONALDS' staff house at the Frost Centre that eight of us — Tom Adler, Penny Clarke, Jim Greenacre, Bill King, Grant Linney, Fred Loosemore, Craig, and I — gathered on February 27, 1982, to get organized for what would be the most ambitious research trip yet — a snowshoe trek along the shores of James Bay from Fort Albany to Moosonee. Craig had been learning more and more about the traditional winter camping technique that he was re-pioneering, based on his interviews with Indigenous informants in n'Daki Menan and beyond, but he wanted to head north into Cree country to see what the limit of these methods might be as the boreal forest gave way to the open expanses of taiga. And to do that, he'd again put out a call for expedition members through the Wilderness Canoe Association.

Since travelling with Craig on the Kipawa-Dumoine trip in 1970, I had corresponded occasionally with him but had been busy getting my own life and research agenda together. By the fall of 1981, I was a junior faculty member in the Outdoor and Experiential Education Unit of the Queen's University Faculty of Education, and when I heard of Craig's plans, following him down the coast of James Bay in the dead of winter seemed like a great idea. The proposed trip would coincide with my undergraduate students' winter practicum, which provided time for faculty members to undertake their own research and to write, unfettered by lectures and classes. I was

From left to right: Jim Greenacre and Penny Clarke (kneeling), Craig Macdonald, James Raffan, Bill King, Tom Adler, Fred Loosemore, and Grant Linney (standing).

fascinated by what Craig was developing with this hot camping technique, particularly as it might relate to doing curriculum-related outdoor camping with students.

On our Kandalore canoe trip together in 1970, I had learned a little bit about Craig's interest in Indigenous knowledge, but back then, as a fifteen-year-old, I had other things on my personal agenda — like growing up — that rendered me more or less oblivious to the nuances of Ojibway place names and a guy who was chasing that kind of knowledge to the ends of the earth. Twelve years later, however, I saw Craig's work differently and was keen to learn what he was doing and how his research was going.

I have saved my journal of that winter trek, along with pages of notes from telephone conversations I had with Craig in the fall of 1981, scratched in pencil on yellow Queen's University memorandum pads. These calls amounted to a crash course in the passions of Craig Kennedy Macdonald. I'd met Doris at a Kandalore New Year's celebration at the camp and knew that they had married in 1973. And I knew that he had taken a job with Lands and Forests and was now working out of the Frost Centre, just up Highway 35 from Kandalore. What was news to me, as we reconnected by phone in anticipation of my joining the 1982 James Bay Expedition, was just how politically involved Craig had become with the Teme-Augama Anishnabai, who were led by Chief Gary Potts. The caution they had filed against the Ontario government in the same summer that Craig and Doris were married had resulted in the Ontario government suing the band to test these claims once and for all before the Supreme Court of Ontario.

My notes from those calls detail what amounts to a comprehensive history lesson about the relationship between the Teme-Augama Anishnabai and the Crown. The details of this evolving and often toxic relationship were buttressed by the addresses and phone numbers that Craig passed on for each of the main players in this drama — people like historian Don Smith, who was Grey Owl's biographer; Bruce Clark, who was head lawyer for the Teme-Augama Anishnabai; Indigenous historian Jim Morrison; information officer Mary Laronde from Bear Island; outfitter Hugh Stewart; and lawyer Dick Grant and his wife Vicki McKenzie, whose grandfather Donald McKenzie had been a key informant for Craig. In Craig's inimitable teacherly way, the notes also record that the plural of "Anishinaabe" is "Anishinaabeg"; that *gama* as a suffix in Ojibway means "water"; and that he was learning these nuances of language from an Anishinaabemowin teacher at Nipissing College in North Bay.

Oh, and by the way, it was the Murphy Lumber Company that built the dam to log Diamond Lake and the B.A. Power Company that, in 1925, needed power for the silver boom in Cobalt that floated Wendawban's house on Lady Evelyn Lake — the same Wendawban who used to grow potatoes in the sandy soils of the Garden Island, now owned by Kirk Wipper. Talking to Craig on the phone was, and is, like getting on a speeding freight train that is accelerating faster and faster. Unless you look to the horizon for the big picture, it's often difficult to know exactly where you are or where you're going.

The big picture was that the Teme-Augama Anishnabai, after registering the land caution, had counter-sued the Crown for malfeasance in matters of their rights and their claims to n'Daki Menan. They were headed to court very soon, and Craig, with all the knowledge he had been gathering about winter and summer routes through the area, was going to be called as a witness in the upcoming trial. A direct quotation from Craig in the notes he made of one phone call: "They have had effect on us. We have had effects on them. But there are certain aspects of their culture that should be preserved."

From his face-to-face encounters with many Bear Islanders, he had a sense that some amongst the Teme-Augama Anishnabai were "demoralized people without a sense of history." What they were getting out of documenting their *nastawgan* with this interested white guy was a "sense of identity." Some of the Elders to whom he spoke, he recorded in his notes, "were proud that somebody took an interest in them and their traditional way of life."

Mixed in with the minutiae of Craig's reporting on the activities leading up to the Teme-Augama Land Claim trial were all sorts of equipment lists and logistical details about the pending winter trip down James Bay's western coast. These were supplemented by phone calls, but the information itself would arrive by mail in classic Craig Macdonald format, that being photocopies of pencil on foolscap. "Dear Participants," he would begin, before recounting the kind of physical activities we might expect, the way the days would roll out, how the tents would be pitched, and who would pay for what and how.

When everyone assembled at the Frost Centre, Craig just assumed that the seven of us had read and taken to heart all of his pre-trip instructions, so the basement briefing began with a review of the food system, tents, stoves, and sliders we would be using. As everyone was heading off for a final pack and bed, with a plan to leave early in the morning, I asked Craig if he might show me what he was working on.

He flipped the switch on a giant light table, and the unfinished basement room glowed red with a layer of film, called Rubylith, used in the printing process — a transparent red acetate on which he was using micro-forceps to remove tiny bits of overlay. As he'd learned about cartography and how it might relate to his desire to create a hard copy of the information he'd been gathering, Craig had found the skills and patience required for the exacting work of creating essentially a stencil, which would be used as the base layer for his map. He didn't even want to guess how many hours he'd spent so far, but just by the way he handled the micro-forceps and the Rubylith itself, it was clear that this was a labour of love.

We set out the following morning for the two-day car-train-plane journey to Fort Albany, located at the mouth of the Albany River on the west coast of James Bay. This was really the opening for me of an illuminating window into Craig's world. Time in the car, time on the train, time on Austin Airways flight 201 from Moosonee to Fort Albany, and time on the trail were all opportunities to chat with Craig about his research and what he'd been learning.

My favourite times were lying in the tent after the day's work was done — and to be sure, this hot camping involved a tremendous amount of effort to find and cut seven poles, and to set up and take down the tent and stove complex, to say nothing of cutting a mountain of firewood to keep warm through nights that dropped below -40°F in many instances. But those times, settling into our sleeping bags on a bed of fresh evergreen boughs in the room-temperature ambience of the tent, after one of Doris's delicious and filling dinners, blowing out the candles and chatting by the red glow and quiet hiss of the tin stove — those times were magical. Sometimes we'd gossip and reminisce about Kandalore. Sometimes we'd wonder about things we'd done or seen that day. And sometimes Craig would tell stories he'd heard from his informants. Other times, in the name of our continuing education and indoctrination into Craig's world, he would talk about stuff he'd read or mentors who had caught his attention along the way.

Not least in Craig's pantheon of wilderness heroes was Québécois coureur de bois Paul Provencher. Some of us were feeling the strain of dragging loaded sleds over James Bay ice, which, when it was very cold, had all the slipperiness of sandpaper. This stressed tendons, particularly in the midsection. When Craig got wind of this, he laughed and diagnosed "Provencher's crotch," a condition brought on by doing heavy hauling labour on snowshoes. And that initiated a story about the old trapper teaching army cadets how to snowshoe as part of their winter survival training. When Provencher's charges started complaining about the very same muscle pains in the midsection, he prescribed a cure that involved cutting a forked stick and putting each tine of the *Y* down a pant leg and letting the third stick ride up over the midriff, over the undershirt but under the outer layers. Of course, said Craig, this did absolutely nothing. It was just Provencher taking the mickey out of these soft young recruits.

On another night, we heard about Dr. John Rae and his days with the Hudson's Bay Company in Moosonee. And how, at some point prior to his heading north to Hudson Bay and eventually finding the key to the Northwest Passage and the unpopular truth of how Sir John Franklin and his men came a cropper, Rae felt the need to travel from Moose Factory to Fort Albany in the winter, covering the hundred-mile distance, which we were planning to do in ten days, in one long day, on his snowshoes. Lying there in the dark, Craig marvelled out loud about Rae's physical acumen, but he also lauded him as an exemplar we should remember because of the way Rae, unlike John Franklin, adopted the dress, techniques, and technologies of his Indigenous friends and allies in the fur trade.

On yet another occasion he impressed us yet again, not only with the exploits of a historical figure, but also with his ability to recall information he'd absorbed from the archival accounts of epic snowshoe journeys. Retelling this tale to me in a later conversation, Craig made clear why he felt that Duncan Sinclair is a name we should remember if we aim to be informed about the length and breadth of achievements in winter camping:

> In many cases, I was looking at more than just the maps. I was looking at survey records of townships but also of these expeditions. There's some fabulous information there. One of the ones that is outstanding to me was the work of Duncan Sinclair when Canada came into being in 1867. The British government decided they should include Rupert's Land in Canada, so they decided to give Ontario what is now the north part of the province. Canada only extended to the height of land; technically, the HBC through their charter controlled all waters flowing into

James Bay. So, in 1867, they appointed some gentlemen to do a survey to find out what this was going to be like.

They had three survey parties. One that went up the Ottawa and up the Montreal River System, and that was run by a fellow named Forest. They did some beautiful mapping of the Montreal River, excellent field notes. Then they had another one from the top end. At a certain parallel, he walked from the Matachewan area westward. And then another individual called Salter started on that same parallel from the area around Wawa, Lake Superior, and went eastward and the idea was for them to join in the middle.

Duncan Sinclair's report is sort of gripping because he went through terrific hardship in northern Temagami. They basically had a surveying party with line cutters and people with instruments, rodman, chainman, and all that. But it was so difficult and so remote, they had to have a supply party, so half the party was involved in that. They got into separation, starvation, and real privation. And then they got trapped by winter and they had to make sleighs and toboggans. They did survive.

I can remember the account of them walking to Fort Mattagami in their duffels, no moccasins, because it was glare ice and they couldn't get any grip. They desperately needed food. Why I found those reports particularly interesting: because they'd make comments about Native people. They substantiated comments that were made to me by a Native Elder in Temagami about land use practices by the Natives. It talked about the creation of rabbit lands through burning, which I found quite interesting. And also it talked about the fierceness of the fur trade competition. What was happening to the fur-trading patterns due to the fierce competition. It was like a game of checkers with these fur trade posts. He documents some of that in his records.

On still other occasions, Craig would talk about his work in documenting canoe routes for the OMNR. He once told us how, in the archives, he found the incredible maps of Augustus Jones and much more besides, including those of the famous historical surveyor and map-maker David Thompson. In this story, again recreated in a later conversation, it was clear that the most significant map of all in Craig's life was the one inside his head, which invariably left the rest of us boggling (and lost) as we followed the historical journeys Craig described:

> I looked at David Thompson's routes and Briscoe and Walpole [Royal Engineers who also surveyed the area] and many more. And I did that for Temagami as well. I looked at those 1837 maps. There was Lieutenant Hawkins and there was David Thompson. A very interesting story. And one thing that was a quest of mine, that I would like to learn more about, which I have not succeeded, although I've gone through Thompson's records, is I'd like to learn how he built cedar canoes. Because out west he was hard pressed to have suitable birchbark, and he learned how to build cedar canoes very rapidly.
>
> For example, in his 1837 expedition through Muskoka Oxtongue and down the Madawaska system, essentially what happened was that after the War of 1812, the Duke of Wellington was very concerned about Americans attacking the St. Lawrence and cutting Upper Canada off from Lower Canada. So he instructed the military engineers from Kingston to go and get Native guides and have them go through trial runs through different routes to see about the feasibility of a canal. Eventually they settled on the Rideau, of course, but they did look at the Magnetawan through Algonquin Park connected with the Petawawa. They looked several times at the Muskoka system connecting with the Madawaska system, and they had different trial runs through there.

> They also knew the French/Mattawa, but then they had David Taylor go even farther north. He turned up the Sturgeon River, went through Lake Temagami, and then down the Matabitchewan and into Temiskaming and did a trial run through there. And the last of the surveys were done in 1837. The Duke of Wellington was following this all very closely and got quite angered at certain individuals for doing things that they shouldn't have done, which means they got guidance from the Indians.... Briscoe got taken to task because — which infuriated the Duke of Wellington — because some Indians at Two Rivers had told him that it was easier to take the portages over to Opeongo, that we did in the winter of 1993, to go that route and then go down through the system, avoiding where he was supposed to be going, which was the Madawaska all the way from the headwaters.
>
> That was one of the reasons why David Thompson was put back in there. But there had been previous attempts. Walpole had gone up the Gull River system up to Minden and then taken the portage up to Kashagawigamog, Drag, Miskwabi, and then got over through Farquhar, Grace, into Elephant Lake and that, of course, is the headwaters of the York River, which puts you on the Madawaska. I took the time to go through the archives and research this.[1]

Craig on the trail was the most unlikely winter travel aficionado, particularly when it came to clothing. While the rest of us worked to choose gear that would acknowledge developments in manufactured fibres and layering techniques, particularly for the purpose of keeping moisture — the bane of a winter traveller's life — at bay, Craig was convinced that from socks to singlets, underwear to outerwear, all-natural cotton, wool, and leather were the only way to go. Though his outdoor work with OMNR, he had made some small nods to progress with his choice of insulated rubber-bottomed, leather-topped boots for work in wet snow conditions, and he did have an old nylon ski jacket that was full of holes, which he said, had been burnt through by acid that spilled from helicopter batteries as he transferred them indoors during winter survey work. But if Craig was about anything when it came to backcountry couture, it was function, rather than form and certainly not colour or style.

The same went for Craig's design of travelling hampers for all the food Doris had prepared for the journey. He reckoned that if we were to use wooden wanigans or, heaven forbid, Tupperware containers, then we'd still have to carry them with us when they were empty. Instead, having had such success with the varnished cardboard tube he used to carry maps on Lake Superior back in 1967, and maybe by now having travelled with enough Indigenous trappers to know that simplest is usually cheapest and best, he created what he called "cardboard wanigans." He chose some big toilet paper boxes and cereal boxes from the stack just inside the grocery store in Huntsville, took them home, and varnished the daylights out of them with multiple coats of Varathane. Then he loaded them up with dry and frozen foods, ready to be lashed to the sleds. When one of these was emptied, he reasoned, we could use it to redistribute loose items we were carrying, or we could just cut it up and use it for kindling.

What became evident as we made our way south from Fort Albany along the shores of James Bay was that for Craig this was not so much a journey on a route from one place to another but rather an exploration of a network of winter trails spanning the country. As we walked and talked, Craig would pull out names of obscure fur trade posts from the map in his head and explain that in the summer, post correspondence was carried from place to place by the canoe brigades. But in the winter, post-to-post mail — the packet — was carried by lone travellers, sometimes with the aid of dogs but often just by hand hauling toboggans and bivouacking en route. And, he explained, if we were thinking that a scant hundred miles between Fort Albany and Moosonee was an epic challenge, we should try some of the longest winter mail routes on for size.

The longest run he'd read about was the Mackenzie River packet from Edmonton to Fort McPherson — 2,012 miles. Then there was the Peace River packet from Edmonton to Hudson's Hope, a distance of over a thousand miles. For the York Factory packet, which carried mail from Winnipeg to the

HBC's original post at the mouth of the Hayes River, a distance of more than seven hundred miles, dog teams would carry the mail north from Lower Fort Garry to Oxford House, and then individual packeteers would hand-haul it for the remainder of the distance to York Factory. And, he added, in the deep snows inland from the shores of James Bay, where we were now, dogs and horses had been no use, so the mail was always carried by hand, and by some of the toughest characters ever to be in the employ of the fur trading companies.[2]

From that fertile mind and memory came more stories of epic winter journeys, always rich with plausible detail. Forcing us to pull from deep in our minds the vague outlines of high school history lessons, Craig reminded us that at the height of the conflict between the North West Company and the HBC, all was not well in Lord Selkirk's Red River Settlement. A free trader called John Pritchard, formerly of the North West Company, was in Montreal and heard from old Nor'Wester friend Donald McKenzie that the North West Company was determined to destroy the HBC's colony at Red River. One thing led to another, and the next thing John Pritchard knew was that Colin Robertson of the HBC in Montreal had engaged him to take news of this deteriorating situation to the Red River Settlement.

Pritchard and a guide set out by canoe from Montreal in October of 1814 and travelled up the Ottawa River and over the divide to the Abitibi River. There, because of the onset of winter, they switched to snowshoes and toboggans. They made their way to Moose Factory, arriving on Boxing Day. It took them only four days, after some rest at Moose Factory, Craig explained, to travel to Fort Albany. We were doing the journey in the opposite direction but taking two and a half times longer. Pritchard's original plan was to hire a guide at Fort Albany to take them up the Albany River to Osnaburgh House and then on to Red River, but that help never materialized. Instead, they followed the icy shores of James Bay, as we were doing, and on up the coast of Hudson Bay to York Factory, where they picked up the packed fur trade trails from there to Red River.

It's one thing to read a story like this, but it's quite a feat to remember it and use it on a daily basis in casual conversation.

Craig went on, telling us that an equally epic snowshoe journey was launched out of the same packet of news the following year. When an attack by the Nor'Westers on Red River looked imminent in the autumn of 1815, a trapper and hunter called Jean-Baptiste Lagimodière, who had been engaged by the governor of the Red River Settlement to provide settlers with buffalo meat, was asked to carry news of this looming conflict to Lord Selkirk himself, in Montreal — in the dead of winter, by snowshoe!

This time, instead of the great north arc that Pritchard had employed, the route they chose was far more direct, thanks to the Indigenous guides they had at their service. Apparently, Lagimodière left Red River in October 1815 with a Cree guide and another HBC operative and travelled east through the Rainy Lake district to a post at Fond du Lac on the St. Louis River, not far from the western tip of Lake Superior. From there, they picked up the Chemin du Fond du Lac, which was an old Indigenous trail that followed the south shore of Lake Superior some 450 miles to Sault Ste. Marie.

By January, Lagimodière and his guide had made it to Drummond Island, then across Manitoulin Island, and along the north shore of Lake Huron. They eventually reached York (present-day Toronto) on March 1, and from there, mercifully, they were able to travel the last leg to Montreal by horse-drawn carriole. When Lord Selkirk again visited Red River a couple of years later, Craig told us, he remembered Lagimodière's services and granted him land on the east side of the Red River, opposite Fort Douglas.[3]

It was as if everything we did on the trail was somehow derivative of or written into some broader story in Craig's head. He had talked to many Elders and trappers about winter travel, the minutiae of snowshoe construction, hand-hauling techniques, and all aspects of camping with a tin wood stove in a canvas wall tent. He had critiqued the manual on the *keejab* trail stove that he had written for the winter camping workshop at the Frost Centre in 1977 after every outing, and just prior to our heading out on the James Bay trek, he had done a major revision of this document. He wanted to write all the learning he had accrued on the trail into instructions that others could use to access the national network of winter trails, or *bonkanah*, on their own.

Jim Greenacre and Grant Linney had been with Craig on the Trans-Algonquin expedition two years previously, so they could help instruct the rest of us in the set-up of the tent and stove, with Craig acting as floating supervisor. It was something of an exacting science. First we had to cut and trim six fifteen-foot poles for scissor and side supports, and one eighteen-foot lanky spruce or balsam pole for the ridge. Craig would remind us to pack down the snow where we were planning to put the tent and to lay the poles down there, to avoid unnecessary extra work. But, whatever you do, he told us, make sure the extension of the ridge pole, the part of the tent structure to which the stovepipe will be wired, is facing downwind.

From there, anyone who was not getting water, unpacking sleds, or collecting firewood would lift and set one end of the tent and then do the same for the other. At that point the corners could be pegged. In many cases on this trip we were setting up on at least three feet of packed snow, so long pegs were in order at most sites. The eave poles would then be lashed to the scissor poles and the tent would take its proper shape. At this point the crew would split in two. Half would put the tarp over the ridge and tie it to the eave poles, and half would go inside and dig out the front third of the floor to set up the stove.

After his early experiments using inefficient trash burners for heating and cooking inside the tent, for which the stovepipe had to be carried separately, Craig had quizzed some of his informants about the dimensions and other granular details of the tin stoves they remembered from the old days. Armed with this information — which included key concepts like the fact that it is the oblique distance between the draft below the door of the stove and the middle of the outlet for the pipe that actually determines how much heat a tin stove will be able to generate — Craig sourced a tinsmith who started building custom stoves for the winter travel innovator.

One of the most important features of Craig's *keejab* trail stove was that, ingeniously, six lengths of pipe, including one piece with a damper, could be tucked inside the stove for transport on a toboggan or sled. A second innovation, developed through experience and through talking with old-timers who knew about these things, was a pair of floorboards or "fire trays" with raised edges for the bottom of the stove. When properly installed, they would prevent hot fuel woods, such as seasoned tamarack, from burning the bottom out of the stove. When cool, these could be inverted with the edges facing upwards, allowing just that much more room for the sections of pipe that would tucked in above.

The idea behind digging out the first third of the tent enclosure was to create a place for the stove to sit on skid logs as well as to create a "frost trough" into which cold air could flow, when the bubble of warm air created by the stove had settled nicely over the bough-covered sleeping platform made from piled snow in the back two-thirds of the tent space. This trough would also allow the stove to draw cold air from outside into the fire, instead of using precious warm air from inside the tent, which was keeping the tent dwellers cozy and drying all manner of sweaty clothing hung on long lines strung along the ridge pole.

Whether the names came from his informants or from his head, Craig had a lexicon of terms to describe all aspects of this hot camping set-up. If our eave poles weren't high enough when tied to the scissor poles, we propped them up with a "spring" pole. The lengths of log on which the stove feet were tacked with common nails were called "skids," and these had to be covered with either a piece of tin or tinfoil to stop them from burning. The lengths of wood cut to the depth of the frost trough and placed to stop the edge of the sleeping platform from melting (causing the sleeping packeteers to slide into the trough and potentially burn the end of their sleeping bag on the side of the stove) were called "billets." The piece of tin or woven asbestos through which the stovepipe passed in the front wall of the tent was called a "thimble." And to thwart Craig's biggest fear — the tent catching fire — we needed to cut "pickets," which were long wands of alder or willow that we stuck into the snow next to the canvas wall of the tent to keep the canvas from touching the stove and to stretch it out a little bit to avoid overlaps or creases where the heat of the stove could fester.

In the early days out from Fort Albany, we stopped in plenty of time to go through this whole set-up procedure while we had daylight to see what we were doing. Later on, however, when we had to lengthen our days on the trail to keep on schedule, the set-up began as the sun was setting and continued well into the darkness.

We experienced a full moon with incredible northern lights later on in the trip, but on one of those earlier nights, with full cloud cover, working in the more or less pitch-dark at -40°F or so, I clearly remember Craig's instruction to "feel the trees" we were cutting for firewood to ensure that there was no bark on them. This would ensure that they were dead, which meant that the wood would burn better, but also that it would not produce gases that would cool and condense in the stovepipe and run back as creosote that would drip onto the stove while we were sleeping and asphyxiate us all!

On another occasion, the wind shifted in the night and started blowing smoke back into the chimney — never fun, always potentially dangerous. Craig matter-of-factly showed us how to split out a flat board and wire it to a thin pole to serve as a wind deflector until the wind changed again or we were able to reorient the chimney, another trick he'd learned somewhere from someone during his interviews.

It was in these little side stories that we fellow travellers got a glimpse of the length, breadth, and depth of Craig's knowledge of Indigenous ways. We were essentially following the same "road" along the coast of James Bay that people like John Pritchard and all of the locals had been following for generations. From time to time, Indigenous travellers moving north and south on their snow machines would come upon us. We would say hello and attempt some pleasantries, but invariably Craig would say something in Cree that would launch a conversation about travel conditions, routes, wildlife, place names, or whatever. I don't think Craig was anywhere near as fluent in Cree as he was in Ojibway, but with key words and a lot of gesticulating with his hands, they had productive exchanges.

In one instance, a Cree guy on a snow machine had stopped and was checking out our sleds and toboggans with great interest. I'm not really sure how the topic came up, but at one point in the conversation he dug into his kit and pulled out a wooden device he called a "snow spoon," which was used for reaching through recently fallen light snow to the deeper water-rich crystals in the snowpack below to make tea. This apparently was new to Craig, who quizzed the guy fully about how it was used and under what conditions, whether there were names for the different types of snow, and so on. This led, as had previous conversations with Indigenous people on the train on the way up and in Moosonee as we were preparing to fly to Fort Albany, to more questions and answers, which Craig duly logged in the winter travel files in his head. To my knowledge, he didn't keep a journal on this or any other trip, beyond making the odd note on the topographic maps he carried amongst his personal gear.

It was clear, however, as we made our way along the shores of Hudson Bay, just how important the conversations Craig had with Indigenous people were for his ongoing research. The central focus of this work was the place names and travel routes of the Teme-Augama Anishnabai, whose land claim was very much on his mind. The land caution registered by Chief Gary Potts on behalf of his community just four days before Craig and Doris were married back in 1973 had niggled away at him. On the trail and chatting at night in the tent, it was evident that Craig's documentation of the *nastawgan* of n'Daki Menan was, in his mind, incontrovertible proof of the Teme-Augama Anishnabai's claim of special ancestral attachment to the lands on his map.

Bruce Clark, the high-energy lawyer for the Teme-Augama Anishnabai, had spent many hours of many days at Craig's house, quizzing him about the nature and veracity of the knowledge he had gleaned from his conversations with Elders on Bear Island and beyond. Craig knew that he was going to have to take the stand in the Supreme Court of Ontario once the trial was under way. This was going to be a very public inquisition, and just the thought of it made him profoundly uncomfortable.

Craig had read the historical accounts of who was where, when, in the evolution of the population of northern Ontario. He'd read the treaties and he was convinced that the people of Temagami did not fall under the conventions of the Robinson-Huron Treaty of 1850 and that they were getting a raw deal. Through both his interviews and his personal travel explorations, he knew, perhaps better than anyone, the preferred routes; where families originated; and when, how, and why they moved from place to place. And, in his heart of hearts, he knew that the knowledge he was transferring from the heads of Elders like Pete Albany, Donald McKenzie, Walter Becker, Madeline Katt Theriault, Andrew Couchie, and June Twain to the map in the basement at the Frost Centre established a long-standing connection between the Teme-Augama Anishnabai and their traditional lands surrounding Temeeaygaming.

But Craig, through his interviews, his work, and his travels on the land, had also witnessed the erosion of language, a diminution of pride, and the decay of a way of life for succeeding generations of people living on and

Craig was able to capture a way of life in Temagami that was in serious transition. But more importantly, he was able to earn the trust of many of the old-timers, like Donald McKenzie, who told Craig things they were not teaching their children and grandchildren because they didn't think it would serve them well in a modern world. Back row, left to right: Helen Twain, Pete Misabie, Catherine (Wassioshic) Misabie, Lucy Pishabo, William Pishabo, Annie Whitebear, Donald McKenzie, Rosie McKenzie. Front row, left to right: Baby, Violet Twain, Susan (Twain) Misabie, Eva McKenzie, Edward Chee Chee. Taken in 1922.

around Bear Island. There was a purity and almost a naïveté to the way he candidly spoke to us on the trail of a sadness in his heart because many of the people he'd interviewed were members of the last generation of Teme-Augama Anishnabai who had personal experience of an on-the-land life infused with the skills, language, place names, routes, and all of the other knowledge that defined a people, a culture, and a land-centric way of life.

It wasn't that his work focused on stopping this decline, or that he saw himself in any kind of heroic light in documenting the ways and the whys of Indigenous land-based learning, but there was something profoundly sad that powered Craig's drive to find his way to the last holders of this knowledge, to capture what he could before it was gone with the people who would otherwise take it to their graves. Craig knew from his association with Bear Island and through Bruce Clark that Gary Potts, the young charismatic chief who filed the land caution and who was now driving the case before the courts, did not speak the language. Gary's experience on the land was mostly localized to his own trapping grounds, so large portions of the land claim area were largely *terra incognita* to him. And with some of his early informants already gone, Craig felt a real and heavy responsibility to have those voices represented at the trial to speak to the power of a dying way of life.

This journey along the coast of James Bay to test the limits of the hot camping techniques Craig was reviving was confirming their viability even in this biological transition zone. The existing winter road was cut inland, well out of the tidal zone of the bay, but even following that, we would have to turn up creeks and move west and inland to ancient beach ridges where better soil drainage allowed the growth of decent-sized black spruce, balsam, and tamarack trees to provide poles and fuel wood. Carrying on, as John Pritchard did to Fort Severn and on to York Factory, would have been a very dicey proposition because of the lack of trees, unless a packeteer was prepared to have a much less elaborate shelter system and to eat cold bannock between the sparce copses.

At lunch on the open ice of the bay one day, Craig took a bite of his frozen peanut butter and jam sandwich, went to his sled, and started rummaging in the cardboard wanigan containing tools, ropes, and other sundry repair items. By and by he produced a carpenter's crosscut wood saw and, as if he'd done it a hundred times before, he nonchalantly started sawing big blocks out of the wind-packed snow at our feet. In no time, he'd created a hole down to the frozen surface of the bay, but he had also carved blocks sufficient to build a wind barrier that looked, from the back, like a classic Inuit iglu under construction. This came with a few pearls of Inuit trail wisdom including a few of the things he'd learned about making *qamatiqs* (Inuit sleds) run well on ice and why he had a quantity of freeze-dried runner mud on hand in the supply cupboard in the basement at the Frost Centre. "Of *course* you have freeze-dried runner mud on hand," a few of us said in unison. We knew Craig had a bit of a mail-order business for tin stoves and canvas tents, but what we didn't know was the impressive range of other products he carried!

As we worked our way down the bay ice, the farther out we travelled to walk a straight route that avoided the undulations in the shoreline, the more the land all started to look the same. As traditional travellers had done for centuries, we estimated our speed of travel and hours spent walking, which allowed us to have a rough idea of where we were on the map — no consumer-grade GPS units were available in 1982 — but still we had a bit of a time trying to locate some of the known landmarks. One of these was Skunk Creek, for which Craig had a particular affection, though I don't remember why.

After about eight days on the trail, we cut back inland to the winter road for the final push back to Moosonee. Jim Greenacre, the sixtysomething member of the expedition, had had enough of the cold by then. Now pulling the lightest toboggan — instead of rotating loads as we had been doing all the way along to share the work of pulling the heaviest sleds — Jim put his head down and more or less disappeared over the horizon, hell-bent for a hot shower at the Lilly Pad BnB, where we'd stayed overnight in Moosonee on the way north.

Craig was peeved that the group was now split, in case something went wrong. Staying together, with the group travelling as a whole, was one of the very few rules he had imposed on the organization of the trip, knowing that unless something very unusual and probably catastrophic happened, there would be safety in numbers. But, like the rest of us, he eventually just shrugged at Jim's bolt, put his shoulder into the load following him, and plodded along, step after step, as we worked our way toward the lights of a microwave antenna on the edge of town. About 9:00 p.m., well after dark, we finally caught up with the aging Englishman at the Lilly Pad.

Looking back, I see now, perhaps more than was evident in 1982, what a privilege it was to travel with Craig Macdonald as he exercised his passion for connecting to untrammelled land through the wisdom and teachings of the Elders. In the spirit of that inspiration, I conclude this participant's view of travelling on the edge of the boreal forest with Craig with the text of a letter I wrote to an old friend, Grace Reid, about the trip. Grace, though not a blood relative, filled the role of grandmother or wise auntie and first took me to the wild. The letter was returned to me upon her death.

Box 18, RR#2
Lyndhurst, Ontario
K0E 1N0
May 25, 1982

Dear Grace:
I've been meaning to write to you for ages now to no avail. The spring always seems to be very busy. However, in between canoeing, studying, moving and the rest I've forced myself to sit down this morning to catch up on my correspondence.

During the first two weeks of March, I went on a snowshoe trek from Fort Albany to Moosonee on the shores of James Bay. The trip proved to be unlike any other of which I have had part. The temperatures were often in the -30°–40°C range, the work pulling our gear on toboggans was exhausting and we saw next to no wildlife. Yet, for some unexplained reason, as I recall the venture, the trip's hardships have faded. Something about the journey inexorably conjured up warmth and good feelings in my mind's eye. I had some prints taken from slides and thought you might like to have a glimpse of this ten-day extravaganza.

There were eight of us, all from different walks of life — seven men and one woman. When the journey ended we called ourselves the Skunk Creek Expedition Society after a creek which took a full day and a half to find.

We (seven of us) all joined one man, Craig Macdonald, in his life's quest to explore and document aboriginal travel methods and routes. Craig has interviewed scores of natives to this end and has built replicas of the snowshoes, long toboggans, tents and stoves which were used by Indians and trappers during the fur trade. To maintain communication between outposts and London the Hudson's Bay Company (HBC) ran mail along routes between posts using pairs of hardy Indians or "packeteers." The "packeteers" would move the mail or "packet" with long (10–12') wooden toboggans pulled by leather tump straps. This second photo is a lithograph of two of our party — we called the expedition the 1982 Winter Packet.

We gathered in Dorset and drove to Cochrane where we boarded the Polar Bear Express for Moosonee. This shot shows the apprehension on Tom Adler's (a newly — 24 hours after writing his final bar exam — qualified lawyer) face. Also, you can see our food "wanigans" which were nothing more than varnished cardboard boxes. The two men in the baggage car are watching other members in our party move a toboggan across the platform — with great amusement. One of the trainmen was heard to remark, "Oh no — there goes another million dollar rescue!" The woman in this picture is an itinerant cook fresh from a logging camp in BC and before that a convent in Toronto. Penny Clarke is her name … quite a girl.

We flew from Moosonee to Fort Albany on a small plane operated by Austin Airways. Instead of the usual golf carts and wagons meeting the plane on the airport ramp, we were met by skidoos and komatiks (sleds) and a four-wheel-drive pickup truck.

The local people were quite friendly, especially the children. While in Fort Albany, we stayed at the Oblate Mission, ate at the small hospital and spent as much time as possible talking to the local people. They were intrigued and amused with our travel methods. Most had heard their parents or grandparents talk of hand-hauling (our method of travel) but few had actually used it. The kids wanted to use our frail toboggans for sliding down the steep, snow-covered banks of the Albany River.

By and large, the land was sparsely treed. The best vegetation grew along streams — most often we camped either on or beside a stream. The firewood for our tent

James Bay expedition members Tom Adler and Penny Clarke ponder loading all the food and gear on the Polar Bear Express, perhaps wondering how the team was going carry all of that on the trail.

stoves (Craig has one on his sled) was most abundant near the streams. Initially, we chopped through the stream ice for water but this was either *very* smelly and muddy or non-existent (stream frozen solid). After being shown by a passing Cree man where the most water-rich snow lay, we eventually just melted snow for water.

At 65, this English salesman had more stamina than *any* of the rest of us. If we weren't careful, he would get so far ahead that the rest of us could not catch up until lunch or camping time.

The tents we slept in were canvas wall structures (10'x12') with an asbestos thimble to allow the stove pipe to pass to the outside. Four people would sleep in each tent. Each tent required seven poles, six at 15' and one at

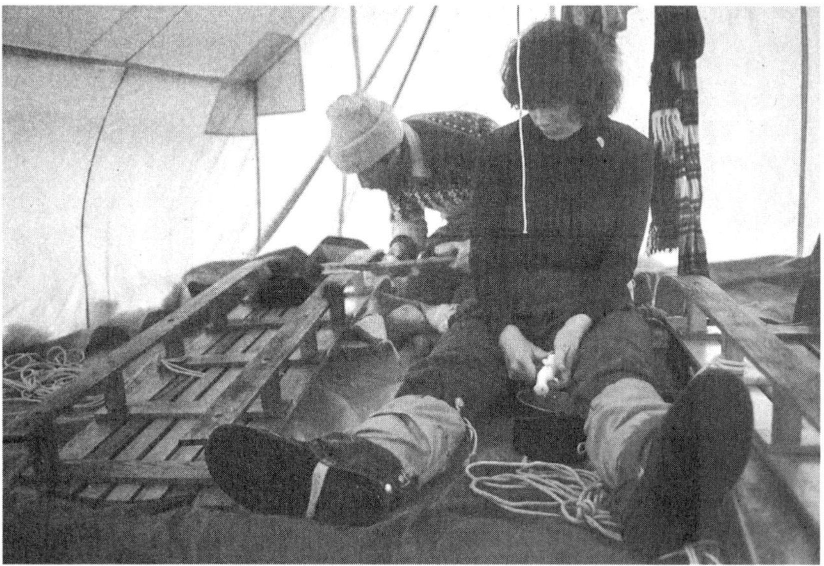

With a heated axe-head, Fred Loosemore irons wax onto the runners of a stanchion sled, as Penny Clarke shaves additional wax from a safety candle.

18'. Needless to say if dead wood wasn't available for poles, cutting live black spruce or tamarack left quite a divot in the landscape!

A view from the front of the tent into the sleeping platform. Here Penny Clarke and Fred Loosemore are waxing the runners of our two sleds. Candle wax would be shaved and then melted onto the birch runners with a hot axe head.

The tents were huge! When the stove was going, the temperature inside the tent could be maintained at 50–60° despite outside temperatures in the -40° range.

Walking on the ice of James Bay was as close as I have been to NeverNeverLand! It could seem endless — the land, the sky, the ice — the walking! Hour after hour with seemingly no progress.

> We had several bad storms on the ice. During white-out conditions we would have to stop often to reorient ourselves.
>
> I think what causes me to look back fondly on this physically punishing trip are the many encounters we had with nature. After a particularly savage storm, we camped at bayside in the dusk. After another exhausting 3-hour setup, we all entered the tent to cook supper. At about midnight, I dressed and went outside. The sky had cleared, the moon was full and the aurora sang like silent wind chimes. I was awestricken. Never before had I had such a feeling that it was only through the benevolence of someone or something that we were allowed to travel in this harsh environment. This photo was taken that night. On the last night of our trip we walked through the pre-midnight darkness to get to Moosonee and the same thing happened. Only this time the aurora covered our trail with an iridescent green arc. Strange.
>
> I should be back in Guelph from time to time. Look forward to seeing you then.

To travel with Craig Macdonald was also to travel with and be fuelled by Doris Macdonald's robust trail cooking. For this journey, she had prepared stews, baked beans, and one-pot wonder concoctions that we took pre-frozen and chopped into pieces to fit them into the pot for heating on the tin stove. But she also provided a range of high-calorie baked goods, which, along with bags of chocolate-rich GORP (good old raisins and peanuts), provided the energy bursts needed to pull sleds through the last afternoon or to cut and split firewood in the dark. Hands down, snowshoeing and winter camping is the best excuse to eat that I've found yet. And of all the comestibles created in Doris Macdonald's kitchen my favourite recipe is her Queen Elizabeth Cake.[4]

Queen Elizabeth Cake

1 cup dates cut fine
1 cup boiling water
1 tsp soda
1 tsp. salt
1 tsp. vanilla
1 cup white sugar
1 egg
½ cup butter
1½ cup sifted all purpose flour

Topping

4 tbsp. melted butter
1 cup coconut
½ cup brown sugar
2 tbsp milk

Pour boiling water over dates, soda and salt. Cool. Cream butter and sugar well. Add egg, beat well.
Blend date mixture alternately with flour.
Add vanilla. Pour into greased floured pan. Bake at 350 for 35 min. After cake is baked spread topping evenly on warm cake and brown under broiler.

Two candles were all it took to light the heated tent on a winter night on the shores of James Bay. But there were often rewards for those who would venture back out after dark to behold heaven and earth.

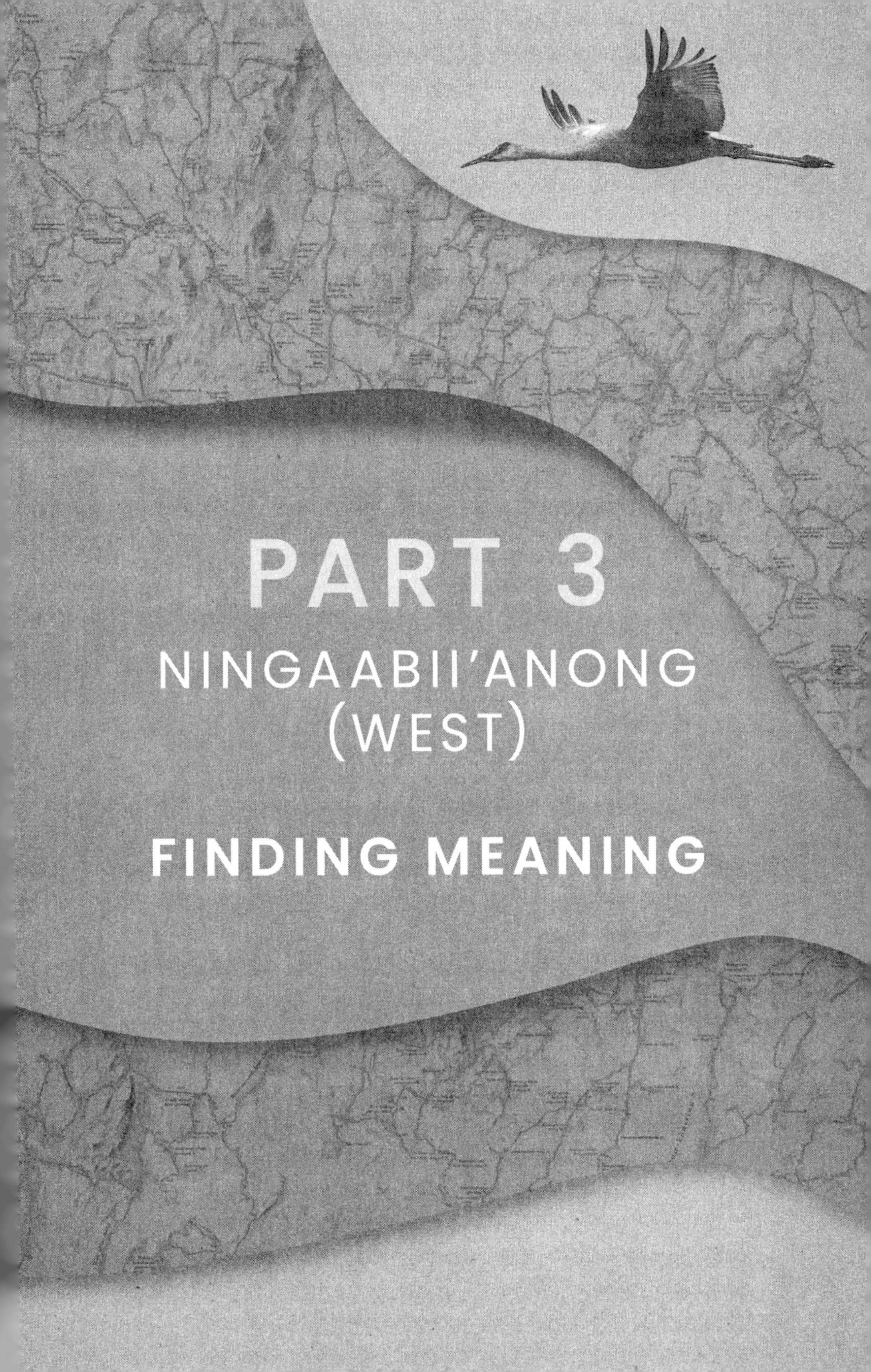

PART 3

NINGAABII'ANONG (WEST)

FINDING MEANING

Mr. Macdonald Is Cross-Examined by T. McCabe, Counsel for the Plaintiff

THURSDAY, FEBRUARY 3, 1983, 10:00 A.M.

His Lordship: Mr. McCabe, are you ready to commence?

Mr. McCabe: Yes, my lord.

Q. Mr. Macdonald, at one point in your testimony you dealt with plaintiff's Exhibit 1-84 ... a map that was prepared in 1880 that purports to show the hunting grounds claimed by the Temagami band at the time.

A. Yes.

Q. Now, what is your information as to the background of the map? Do you know by whom it was made and under what circumstances?

A. No. I would say my information is limited. I solely examined it on what I see in front of me.

Q. I take it then in connection with this map, all you have seen is the map itself?

A. That is correct, I have not read the correspondence.

[Correspondence from the 1880s between the maker of the map, Charles Skene, Indian Superintendent at Parry Sound, and his boss, Deputy Superintendent of Indian Affairs, is read into the record by Mr. McCabe.]

Mr. McCabe [as he finishes reading this correspondence into the court record]: Then, over on the next page of that letter about six lines from the end: "Chief Tonenee [Tonené] also told me that it is along the bound which runs parallel to and a short distance from the Montreal River that trees have been taken. This is a condensed reply of what I heard from Chief Tonenee and other Indians with whom I had several interviews while waiting two days for Chief Cochai and his band, but I think it contains all that was said of any importance."

The Witness: I am confused here. I have lost what we skipped and I don't know what is going on in the letter.

Mr. McCabe: In the last letter?

A. Yes.

Q. I read the first page of the letter, then I skipped down to the point about six lines from the end of the letter, beginning "Chief Tonenee."

A. You are jumping in and I don't know what is going on. What is this business of treaties? Unfortunately I am not good at reading this stuff in handwriting.

Q. So you prefer that I read the entire letter?

A. Cover the part you jumped.

Q. All right [then McCabe reads the letter in its entirety]. So, Mr. Macdonald, this map then, plaintiff's Exhibit I-84, apparently reflects reality as perceived by Chief Tonenee of the Temagami Indians in 1880 and by Gilbert Pawis who knew the country well, we are told, and by several of the Lake Nipissing Indians.

A. Just a second here. First off, I would want to make it clear, I don't agree with what is in there as far as the mapping is concerned, the quality of the map, I have already stated that Alexander Murray in 1854 had an excellent survey that included the Wanapitei Lakes over the Sturgeon River, importance of the Sturgeon River. This is many years before it, they had excellent latitudinal control. I can also say that in 1867 Forrest did an excellent track survey of the Montreal River. Also Duncan Sinclair did excellent line surveys across the top; also side surveys.

To say — I don't know — maybe there was a problem — maybe they didn't have access for some darn reason, but to say — you know, there was good mapping, there was people knew how these places lay.

Q. Are you saying that Skene is mistaken when he said there wasn't good maps?
A. Absolutely.
Q. Apart from the base map that wasn't available to Skene —
A. You read to me some stuff where Tonenee didn't agree with this thing. He said the map was incorrect. That is what I heard you read.
Q. Did you also hear the part where Tonenee marked or agreed to or told Skene as to the boundaries of the Temagami Indians, the Temagami band of Indians?
A. I'm sorry. It's so long. I would have to hear it again, but what I got from the letter —
Mr. Clark: The places in the letter should be pointed out. It is not marked and the witness should not be misled.
His Lordship: I think the whole letter has been read at this stage. Just because I look a little askance doesn't mean you need to jump up, Mr. Clark.
The Witness: This is the first time I have seen this. What I get from the letter, he says the map is incorrect. There may be other things there, but I just don't have them in my mind …
Q. Apparently the base map or some of the geographical materials on the map is incorrect in the view of Tonenee, but he does, on the map, point out the bounds of the Temagami Indians?
A. I have difficulty in — okay. My problem is understanding what sort of validity is being placed on this. He is obviously working through translators. You know, there is no — it is speculation that what is really going on here. He said the map was incorrect. He probably just sort of went "wu, wu, wu," and put his boundary on. I can't speculate on that.
Q. Please don't speculate. The text of the letter, the words used by Skene in describing the events of the 27th of July or thereabouts when he was on the scene, are to the effect that Tonenee pointed out the bounds.
A. On the map he said was incorrect.
Q. All right. Gilbert Pawis, we are told, agreed with Chief Tonenee as to the bounds claimed by the band … and said, "I have no doubt it was as nearly correct as could well be made without an actual survey."

A. Now I have told you that there were paper maps available.

Q. But apparently on this occasion he doesn't have the benefit of those proper surveys and is doing the best he can on the basis of what he has and on the basis of what he has been told by these Indians that he is dealing with.

A. Okay.

Q. ... I put it to you, whatever you may think there of the cartographic merits of the map, the plain truth is, the plain fact is that these three groups, Tonenee, Gilbert Pawis and the Nipissing Indian informants of Skene on this occasion, all concurred that the map reflects the reality of the bounds of the Temagami Indians at that time. Now, again, I am not asking you ... Quite apart from the merits of the map, that is what we are told in the letter, isn't it, Mr. Macdonald?

A. I don't know how to articulate it in a way that — if you are going to mark things — you are talking about agreements and marking things on maps. You are making — you have to make assumptions that the map is reasonably correct and they have told you that the map isn't correct. How could they possibly —

Q. Well —

A. Again, I don't really know what is going on here.

Q. Mr. Macdonald, as a person who deals with maps, who is familiar with maps, who is familiar with the geography of the area, you may well take that view. I think you had very disparaging comments to make about the quality of that map, but what we are told, quite apart from the merits or demerits of the map, there is on that map the hunting grounds, the bounds of the hunting ground of the Temagami Indians appear; is that correct? ...

A. Well, we are at an impasse here, because, you know, I don't know — do you understand what I am trying to say?

Q. I do. I put it to you, that this map purports to reflect a reality as of the summer of 1880. Don't consider at this point the merits of the map.

A. Okay. I am following you now.

Q. Do you agree that this map purports to —

A. It is purporting.

Q. Now, in doing so, this map is not based on a reconstruction of events and circumstances made by a person over a hundred years later on the basis of some interviews in the 1970s of the people who in all probability were not born in 1880; is that correct?

A. Okay. What I can say about that is, if we are getting into the business of this oral testimony, I think if we did some mathematics, let's take for example, let's take for example Donald McKenzie, we know that he died at approximately age eighty-six and this was, I think he died in around 1980.

Okay. If you work it backwards, what would that make him born in? In the 1890s, say he had a conversation with his grandmother or his grandfather, that is very reasonable, assuming where he is getting his information from, you know that it would go back to that time period.

Q. Yes. And at this point, all I am asking you for is your agreement that the information that purports to be set out on this map is not based on a reconstruction many decades later, based on what informants at that time may have been told may years previously; is it?

A. This map was produced at the time period it said, yes.

Q. Now, as between reconstruction of circumstances a century later based on informant testimony of persons who were probably not born at the time of the circumstances in question; and a record made right at the time of the events in question, which, in your view is likely to be the most accurate and again at this point please don't allow your answer to be coloured by your estimation of the cartographic merits of the map?

A. It is bias, I can't help it. I am fundamentally biased. When I look at that and it says that the guys disagreed with it and I can't go further. You know, I don't see what the problem is here.

Q. Mr. Macdonald —

A. I look at the map, just take a look at it, and you ask me what I am trying to think of it — look at the north end of Lake Temagami, what can you relate from that?

Q. Let's look at the map in a few minutes … as between a reconstruction of events a century later based on informant testimony on the one hand,

and a document that purports to record reality contemporaneously with the circumstances that it purports to record; in the abstract, which of those two is likely to be more accurate in telling us about the reality of the period in question?

A. Well, quite frankly, because of the great detail I have gone into, I would trust the oral testimony.

Q. All right ...

A. I can't understand it, really.

Q. But apart from the map itself, apart from the merits or demerits of the map, as an abstract proposition?

A. As an abstract proposition, I have told you I have interviewed a large number of people over a great number of years. These are very senior people. We have gone through this in great gory detail, hill by hill, in cases they have given me all sorts of — I have reams of information we haven't even discussed here ...

Q. You have a lot of information about portage alignments and hills and mountains —

A. Many other kinds of geographic features to the land claim — these lands and I have features that you have seen marked as boundaries. I have given them to you.

Q. I think you testified you weren't interested in boundaries, that is not the object of your investigation?

A. Yes, that is true.

Q. I put it to you, that you are not in any position, even given the admirable wealth of detail that you have recorded ... to contradict the evidence of this document that actually recorded at the time that the actual circumstances recorded was in effect?

A. Let me answer you, that I — this way. That what you are saying is absolutely correct in terms of you have said that my principal thing was geographic information ... and this business on boundaries is purely secondary.... I wasn't interested in land claims. Okay ... I couldn't give a darn about Donald McKenzie serving in World War One. That doesn't negate the fact that he doesn't tell me about it ...

Q. Yes. You testified —

A. Okay. From the outside. You started objecting and you thwarted me from giving testimony concerning living people. This has greatly reduced what I can tell the court. I am sorry, but my opinion is that I don't understand the legal process, but it seems that the cards are heavily stacked against these people, because —

His Lordship: Just a moment. I don't want that to go unchallenged. In the normal action or trial, the cards are very heavily stacked against hearsay evidence which means somebody telling the court something somebody has told them. In a trial involving Indian oral testimony this is permitted. If it was a different kind of trial, you wouldn't have been able to give any of this. We are into it now, but there are limits. I don't think you were told you couldn't tell anything of what living persons said. It is more a question if there is a living person that that living person should be here in the courtroom giving the evidence. If they are deceased, in a trial involving Indian people, you can refer to that. I have allowed you, and indeed you have made several references to living persons, and there wasn't any serious objection.[1]

Lifeline III — The Map Comes to Be

THE FESTERING LAWSUIT THAT WOULD RESULT IN TIMOTHY McCabe's withering cross-examination of Craig in the Supreme Court of Ontario in Toronto had, at least in its early days, a salutary effect on the research he was doing. From the moment in 1973 when Chief Gary Potts and lawyer Bruce Clark had registered the land caution for the whole of n'Daki Menan, and for the first time ever, a semi-public spotlight illuminated the work Craig had been doing in more or less obscurity since the late 1960s. Chief Potts got wind of a guy who was systematically interviewing Elders as well as traditional, recreational, and commercial users of the greater Temagami area, gathering information about place names, summer and winter travel routes, and other details of special places, and it became clear that this body of research was potentially highly relevant evidence for the trial that would eventually happen.

Of course, Craig had a day job and, as of 1976, a growing family, which relegated his research to an avocation at best. But with Gary Potts's help, the co-operation of Doris, and his own sheer determination, Craig was able to slowly ramp up his interviewing schedule so that in 1977 alone, he conducted more than two dozen interviews in the Temagami area. He went back to Mowat's Landing for a formal interview with Pete Albany, and from that flowed other key conversations with people such as Batisse Tonnice at Matachewan Reserve; John Commanda, John Fisher, Fred McLeod Senior,

A pencil sketch of Donald McKenzie by his son, Hugh McKenzie.

and Chief Paul Goulais at Garden Village on Lake Nipissing; Andrew Couchie at the Temagami Inn on Temagami Island (where Craig first heard the *War of the Worlds* story); Aubrey Dunne in Huntsville; Wilfred McBride in North Temiskaming; Michael Kush Kush Paul in Corbeil; Jack Pierce in Elk Lake; Madeline Katt Theriault in North Bay; and George Turner on Temeeaygaming. He even traced Emerson McConnell to the North Cobalt Seniors' Home and Angelique Misabi to the New Liskeard Hospital. Perhaps the most important informant Craig interviewed was Elder Donald McKenzie at Makõminising, with whom he had long conversations in January and March 1977.

Gaining access to these informants was just one step in the process, and Craig was aided by previous informants who passed on to others their good feelings about what he was doing. People like Pete Albany suggested names of others to whom Craig might speak, and in some cases made introductions.

The same with Chief Potts, who introduced Craig to Bear Island Elders and endorsed the work he was doing.

But gaining the trust of these people and asking them to speak on such personal and sensitive topics, this was a whole other matter. Craig's passion to learn and his quiet and unassuming manner helped immeasurably. He wore everyday jeans with suspenders and a work shirt, maybe topped with a grease-stained jacket from his work at the OMNR. This modest presentation, along with Craig's increasing command of Anishinaabemowin and his impressive memory for place names, land features, and the stories that went with them, allowed people to be comfortable detailing routes, place names, and travel technologies used throughout n'Daki Menan.

It didn't seem to matter where Craig did the interviewing: in people's homes, in the Band Office, in cafés, or in old folks' homes. He engaged people with his interest and authenticity and charmed them with his child-like enthusiasm for knowledge that, for many of his informants, had depreciated sufficiently in value in the wage economy that it was now considered almost irrelevant. One exception to this almost universally positive first impression occurred on the day Craig drove up to North Bay and walked into the lounge of the Empire Hotel where, others had told him, he would likely find Walter Becker. Indeed, he did find Walter and did talk with him, but only after he had dispelled Walter's suspicion that this strange guy coming up to him in the bar was some kind of government agent who was there to call him up on past improprieties with his veteran's pension. With Walter's suspicions put to rest, Craig extracted his master map, with all of its pencil notations, from the cardboard tube he carried under his arm and spread it out on the table, and off they went on a journey into Walter's past life on the land.

Craig went about this work in almost total ethnographic isolation. His research had connected him to the early work of kindred historical figures like Augustus Jones, Edwin Tappan Adney, Paul Provencher, William Francis Ganong, Duncan Sinclair, Frank Speck, Irving Hallowell, and others, whose accounts he had read in archives and libraries as he was researching canoe routes. Rather than seeking out contemporary historians, ethnographers, linguists, or cartographers who might have been able to add to his knowledge from their ongoing academic research programs, Craig's singular focus was on gathering primary information from the aging demographic of people who had actually lived and travelled on the land year-round. His brother, Rod, was the lawyer and scholar. Craig saw himself more as the everyman, grounded in the experience of the land and in the perspectives of like-minded practical knowledge holders.

That changed, however, when he got on the stand. The Temagami land claim trial opened in June 1982 with lawyerly fussing that would lead, over the next couple of years, after various scheduling delays, to the Crown putting forward its case and, eventually, the Teme-Augama Anishnabai getting their day in court to establish once and for all their rightful Aboriginal title to n'Daki Menan.

In August 1982, Bruce Clark and Chief Potts convened a most remarkable constellation of academics at Bear Island to put their heads together to prepare a vigorous defence of their homeland. One of those in attendance was Brian Slattery, a brilliant lawyer who had just completed a doctorate from Oxford University on the Royal Proclamation of 1763. Bruce Clark had tracked him down at the University of Dar es Salaam in Tanzania. Next were the heavyweight team of Edward S. Rogers, head of the Department of Ethnology at the Royal Ontario Museum, and his wife, Mary Black-Rogers, a distinguished anthropologist and ethnohistorian.

Also in the group was a former Ph.D. student of Ed Rogers, Donald B. Smith, an up-and-coming historian and author who would go on to a luminous teaching career at the University of Calgary and would become Grey Owl's biographer. He would likely have had copies of his brand new first book, *Long Lance: The True Story of an Imposter*, a penetrating biography of the erstwhile star of the Hollywood film *Silent Enemy*, which lurched Bear Island into the modern era in 1929. Smith remembers being impressed with Craig's mention of his two hundred interviews to date and of his growing command of Anishinaabemowin. And along with Don Smith was another key player in the Temagami saga, a historian friend, Jim Morrison, who with his physician wife, Gretchen Roedde, had moved to Bear Island in the mid '70s to work full-time on building the Teme-Augama Anishnabai's case.

The Bear Island Pretrial Summit, July 1982. From left to right: Brian Slattery, Osgoode Hall law professor; Bill Twain, former Bear Island Chief and respected Elder; Craig Macdonald; Mary Black-Rogers, anthropologist and ethnohistorian; Jim Morrison, historical researcher; Donald B. Smith, author and historian; Bruce Clark, lawyer; Gary Potts, Chief of Bear Island Teme-Augama Anishnabai; Margaret Clark, spouse of Bruce Clark; Jim Wright, archaeologist; W.J. Eccles, historian; Thor and Julie Conway, Great Lakes archaeologists; and E.S. Rogers, curator of ethnology at the Royal Ontario Museum.

Bringing deep knowledge of Canada's French colonial history to the gathering was English historian and academic William J. Eccles, who was recognized as the foremost writer on this topic at the time. Also, there were three archaeologists: James Wright, who had been Chief and Senior Research Scientist at the Archaeological Survey of Canada, and Thor and Julie Conway, who brought to the meeting a trove of knowledge about rock paintings and the history and prehistory of the area.

And then there was Craig. Although he did have a respectable post-secondary pedigree, his invitation to this gathering had nothing to do with his encyclopedic knowledge of the cephalic-lateralis system of the fish family Gobiidae. In this group of distinguished academics, he was more or less a total unknown and might have been considered a rube or jackknife ethnologist — at least until it came time to talk about the map he was working on. In his own quiet way he captured the attention of this ivory-tower crowd, and this event became something of a watershed moment, a coming out into the world at large of his passions and remarkable research.

Interestingly, the only members of the Bear Island community who attended this meeting were Chief Gary Potts and Elders Bill Twain and Maurice McKenzie, who were named defendants in the original case and then plaintiffs ("on behalf of themselves and on behalf of all other members of the Teme-agama Anishnabay and Temagami Band of Indians") by counterclaim in the lawsuit. They floated in and out of the gathering on Bear Island from time to time. Chief Potts and Bruce Clark had decided that the best way to persuade the court of the merits of their claim was to barrage the judge with testimony from non-Indigenous experts. But this strategy would prove problematic, as Justice Donald Steele, when he eventually ruled against the Teme-Augama Anishnabai, asked why more community members were not on the witness list.

The format for the gathering was for each of the participants to present on their particular interests and expertise relating to the trial. These indoor sessions were punctuated with outings on the lake, including to the archaeological dig site at nearby Witch Point, where they were shown flint arrowheads from another site on the Montreal River dating back to 3,000 BCE.

In his first presentation of his work to an academic audience, Craig stunned his fellow workshop participants with the depth of his knowledge. Donald Smith, who by now had done a lot of his Grey Owl research, was totally impressed that Craig had spent time with Jim and Susan Espaniel at their home in Spanish River country west of Sudbury. He was surprised that Craig had first-hand details of an altercation between Grey Owl, who was living in Biscotasing, and Frank Beaucage, a so-called railway trapper who had moved over toward Bisco from his home area on Chiniguchi Lake near Lake Wanapitei when it got too crowded. Grey Owl had taken to Beaucage with a curved monkey wrench, but Beaucage ducked and quickly overpowered the reedy Englishman. But the point was made. Or so the story went.

Craig and Don Smith talked about Grey Owl and their research throughout the time they had together on Bear Island that summer. What became evident was that whatever methods Craig was using to talk to people, he was somehow accessing details that Smith was not. In his journal from the day, Smith noted that Craig told him that Jim Espaniel had volunteered to Craig that Grey Owl wasn't nearly as tough as he presented himself to be (case in point the monkey wrench incident). And Craig said that Jim had painted a much more benevolent picture of the fabled imposter than the one that history has passed down. "He was a good man. Kind," Craig told Don Smith.

Talking to me about that day, Smith said, "Wow! Those are six words which are the redemption of Archie Belaney. 'He was a good man. Kind.' That's gold! That's from Craig. I could write ten pages on that. Because I know Jimmy. I know what he was like. I talked to Jim too but, you know, that direct stuff doesn't work. Craig's got it. 'He was a good man. Kind.' Craig's like a boulder going into a little pond. Poof! And there are the waves, the repercussions that go on for years."

Don Smith was not the only academic whose world was rocked by Craig's participation at the pretrial conference on Bear Island. When the esteemed ethnologist Edward Rogers made his presentation, he stated that it was his belief that there was no First Nations agriculture north of Wasauksing (Parry Island, Georgian Bay). Here's Craig's description of what happened next:

> I says, "Hold on here a minute." And I showed him a treaty map, which each one of these areas had a major garden. And then I talked to him about the fact that other work I'd done indicated that there were cornfields as far up as Lady Evelyn. There was a big corn operation. That just blew him away.
>
> Garden Island on Lady Evelyn is Garden Island because that was Wendawban's garden, I told him. That's where he planted the potatoes. Easy digging because of the sand. The potatoes would grow in the sand. But more importantly, by putting it out on an island, it was less prone to frost.... If it was put inland, you know, you could have an early frost or a frost in June or ... July and ruin your crop.

> He [Dr. Rogers] just took the pages of his report and dumped them in the garbage. He says, "I need to rethink this whole thing. I need to start again." It was symbolic. I'm sure he pulled them out of the garbage later, but he just said, "I've got to do major revisions here." He got a copy of the map and could see clearly that his whole concept had to be altered with regard to gardens.

Of those in attendance at the pretrial conference, no one was more impressed with Craig and his encyclopedic knowledge of the *nastawgan*, toponymy, and lore of n'Daki Menan than Bruce Clark. Clark was convinced that what the judge needed to hear, more than any testimony from Bear Island Elders or even from any of these other experts, was a steady decanting of the nearly five hundred place names and their associated meanings that Craig had learned through his research.

When it was Clark's turn to present evidence to the court, on Thursday, January 27, 1983, he put Craig on the stand, where he testified until Friday, February 4. After an opening walk-through of Craig's credentials and the length and breadth of his research activity, Clark introduced to the court the physical manifestation of Craig's work. Propped up on an easel in the forecourt was the photographic rendering of a provisional version of Craig's map along with a full separate list of geographic names. These exhibits were numbered for the purpose of the trial as Exhibits 19-1 and 19-2. For clarity, each major map feature mentioned was given a number. Then, in groups of related features, Craig presented to the court, over his seven days of testimony, the Anishinaabemowin names for lakes and rivers, relating them to the common English names and detailing the meanings of the traditional nomenclature.

Almost immediately as Craig began his testimony, Tim McCabe, lead counsel for the Crown, jumped to his feet, objecting to Craig reporting on conversations he'd had with people who were, at the time of the trial, still living. If an informant was still living, McCabe argued, they should be called as a witness. And right from the beginning, Justice Steele agreed, saying that in normal circumstances, anything a witness reported that they had heard from another living person would be considered hearsay and, as such, would not be acceptable evidence. Steele did concede that in cases involving oral recounting of Indigenous history, the hearsay rule could be relaxed a little, but he also said that, in the end, he would have to discount the value of Craig's testimony about the traditional use of the land claim area by Teme-Augama Anishnabai, especially if none of his informants who were still living were called as witnesses to buttress his claims.

The court transcript shows that during Craig's cross-examination, the Crown opted to question him on the written historical evidence of occupation of n'Daki Menan, notably some correspondence and a map, circa 1880, relating to the home territory of the Teme-Augama Anishnabai drawn from the records of Charles Skene, the federal Indian Superintendent at Parry Sound. If Craig had seen this map in his research, he had likely not internalized its contents, or perhaps he had dismissed what it contained, because from the moment McCabe started in on this line of inquiry, Craig got increasingly flustered. Starting to sputter, he testified that he had not seen the correspondence referring to comments by Chief Ignace Tonené ("Tonenee" in the trial records), then head of the Teme-Augama Anishnabai, and he couldn't fathom why the Crown persisted with questions about something clearly outside his area of interest and expertise.

In Craig's mind, he had talked to people, like Donald McKenzie, who were born in the nineteenth century and who spoke to him of their experiences on the land in question. Their oral recounting of traditional territories and how they were accessed and used was equally, and possibly more, valid than any written historical record or hand-annotated map based on conversations that may or may not have happened. That was Craig's view. Nothing academic or esoteric about it. As far as Craig was concerned, what he had learned from his informants was the truth.

So when Crown lawyer Tim McCabe poked at Craig until he lost his cool, Craig's notions of truth were badly shaken. It had likely never occurred to him that what people like Donald McKenzie told him was anything but the absolute truth. Craig's life's work was anchored to the principle that what his informants had to say was drawn from their own long experience on the land and informed by generations of stories and accounts of their forebears on the very same trails, and thereby unchallengeable. To have the court suggest that Craig's knowledge, gathered from all his interviews and trips, was somehow substandard testimony — well, that was almost too much to bear.

Craig's daughter Nancy, who was seven at the time, remembers her dad coming home from Toronto that winter. "I think it was a painful experience for him," she told me. "And it's not something that he talks about frequently. It was not pleasant. My dad doesn't show his emotions very often. But he's actually a very emotional person. So to go through something like this, for my dad, is very upsetting. The trial touched him more than you would think."[1]

Daughter Janet, although just four at the time, also recalls the trial. She told me, "A lot of that was held away from us. I remember the stress and the tension in the house. And I remember Dad coming back skinnier. Other than that, it wasn't really discussed in front of us."

Once the trial was over, Craig continued his work at the Frost Centre and doubled down, when he could, on his research and map-making. He could sometimes kill two birds with one stone by taking advantage of family holidays, such as a canoe trip with Doris and the kids in Temagami or road trips to areas where Craig knew they could have some fun but might also provide opportunities to do an interview or two to advance his knowledge of new areas.

There's an epic family story of a road trip in the family Suburban to Lake Superior. Two adults. Three kids under six. Reaching the town of Longlac early one morning, Doris was keen to do some washing, so they stopped at the local laundromat. Craig said he was just going to check and see if Joe Bananish was around, because he'd been wanting to talk to him about his life on the land, which he'd heard connected somehow right up to James Bay. "I'll be right back," Craig said, famously, as he hopped in the Suburban and drove away.

By noon, the laundry was all clean and folded but the little ones needed changing, and Doris discovered that Craig had disappeared with their supply of diapers in the truck. With the kids and the laundry, she left the laundromat and headed downtown, where she was able to purchase some diapers and some food to keep them going as the vigil continued.

Eight hours later, as afternoon had given way to a summer evening in Longlac, Craig hied up and found his family in a little park by the lake. By then, Doris was fuming. Looking back, she recalled,

> I was pretty angry. Yes. Very angry. What do you do? And then he says, "This guy was talking and I needed all his information." But I look back and it's just Craig's personality. He forgets. He's into something and he's so focused that we were likely not even thought about. The trouble is, when you're in the middle of that, you think he's going to return any minute now. Oh, he'll be here in a few minutes. It gets to be twelve o'clock and he's not back. And I'm thinking, *What are we going to be doing about food?* I think I waited until about one o'clock and then finally just said to Nancy (who's five or six), "Look after the kids. You stay with the kids and play in the park. I'm going over to the local store over here and getting this food." I also had two massive bundles of laundry that I'd just done!

Nancy remembers Longlac too:

> I have a memory of the laundromat and that we were there for a very long time. I remember that there was really nothing there on the street. It was boring. You tried to amuse yourself. That's pretty well it. As much as we're talking about the laundromat story, there's actually many stories like this. He also left us under a tree for two or three hours, while he went off looking for a portage. For me the laundromat story is just one more along the same lines of being sort of abandoned for long periods of time in nowheresville because my dad has decided that he needs to go off and do interviews of some sort.

Craig's recollection of that incident went this way:

> For me, it went wonderfully.... I was interviewing Joe Bananish. It was a phenomenal interview. When I got back, she was a little hot under the collar because I had no idea this would morph into an eight-hour effort. But it was well worthwhile. I got a lake-by-lake, trail-by-trail description of the traditional snowshoe route that was used for the packeteers that were going between Michipicoten to Pic River Post to Longlac Post to Nipigon House and out to Fort William. So I know the route across the Otter Peninsula. I know exactly where they went, lake by lake, trail by trail.

Surprisingly, given the ups and downs of the Macdonald family adventures at home and on the road, Doris saw these frustrations as family investments in Craig's work, in Craig's life's passion. This crystallized for Doris when a Huntsville friend of Craig's, one of the organizers of the original winter camping workshop at the Frost Centre, began to spend more and more time at their home, quizzing him and taking detailed notes about his research, his knowledge of trails, and his incredible body of work on Indigenous winter travel technology and techniques. The friend mentioned that he was thinking of writing a book with what he was learning from Craig. Doris recalled how she bristled at that idea: "That time ... Okay, I told him, I've dedicated my life to this project. Basically, what I've done is raised your family for you. I've kept the household going for you. I've looked after all the finances. I've done everything for you. You're not giving this book away now. Or this map away. Whatever it is. We're *not* giving it away!"[2]

There was more than family drama and politics associated with the map and Craig's research. As one might expect, a charismatic chief like Gary Potts, who poked the government bear with his land caution and was now all in on proving Aboriginal title to n'Daki Menan, made as many enemies as he did friends as he and Bruce Clark led the community into litigation. There were community members who were decidedly not fans of Chief Potts. Because of Craig's association with Gary Potts, some Elders who were on his list were perennially "unavailable" for a conversation. Among those were George and Charlotte Pishabo. George's family background was in the Anima Nipissing area, part of the map for which Craig didn't have much good information, but the Pishabos could never seem to find the time to speak with Craig.

Enter historian Jim Morrison and his wife, physician Gretchen Roedde, who were invited into the Bear Island community by Gary Potts to help build the land claim case. Unbeknownst to Craig, George Pishabo had been suffering increasingly from diabetes and, through the diagnostic skills and subsequent medical ministrations of Dr. Roedde, his condition was brought under control. Suddenly, the Pishabos had time for Craig. Here is how Craig described the change of heart:

> I just went to his house and knocked on the door. Hauled out the maps and he was in quite a different mood.... He showed me stuff from his personal life. Photographs of him playing hockey in the 1930s. Talked about his guiding experiences with all sorts of people, including Gordie Howe. And then we got into what I was principally interested in. He was in his mid-80s at this point, maybe 86, I just can't remember now. I was more specific about the questioning. So it wasn't just a rambling thing about his life, it was more specific questions and answers about Shkimskajeeshing and Mink Ball Lake and these places that he had trapped.
>
> And also, one very difficult problem I had was trying to sort out the name for Scarecrow Lake. He knew it at one time, but he couldn't dredge it up in his memory. And he was the only one in the band that knew it. I never did find out what it was. A lot of Native people use *say-gizhi-andeg*, which means scared crow, but that wasn't its original traditional name. What I wanted to do was find this out because I was absolutely frustrated out west with the information because one of the people who really knew the stuff was the Chief Paul from away back, and he had died a year prior to when I started this work. If I had talked to

him, the western side of the Temagami map would have been much different, much more detail. It's thin. It could have been much better. But George was very helpful on the physical end of things, the snowshoe routes. The snowshoe routes, you have to understand, were different.

Talking to George Pishabo, I have the maps on the ground. We're following them. Sometimes, if it was a little bit hard to understand, he would draw for me, but often I could draw myself. I could see what he was talking about and I would do the drawing. On occasion, the person I was interviewing would actually pick up the pencil and draw in stuff. All of them, anybody who had been interviewed, would study the map for thirty seconds. Some of these guys had not used maps, but could very quickly figure out what's going on, looking at the shapes of lakes and stuff like that. They knew, *Oh yeah, it's here, and it's here.* It was easy for them to figure out. In that regard it was surprisingly easy.

Here's an example of how good George Pishabo was, why I treasured what he was telling me. There's a lake up by Anvil Lake that's called Skull Lake. It gets its name from a great big rock that's on the south end of the lake that looks like a skull. I was getting translations from other people, something like *she-bay-tig-wan*, which means skeleton. *Chee-bay* is a spirit or a ghost whereas *she-bay* is a skeleton. So *she-bay-tig-wan* is skeleton head, which could be translated as skull. And that's what a number of people were giving me. But I'd seen an old map from around 1900 done by, I think, the Forestry Branch. I'd seen it in the archives at Grenville and it had a longer name for Skull Lake. So I gave that name — She-bay-tig-wan — to George Pishabo and he says, "Ah, no. That's not right." He says that's *Chee-bay-tig-waniga*. It's the spirit head. He had the longer rendition of the word. I checked again with my historic map and found him to be right. He's the only guy that knew that. So he was a very valuable person to talk to.

Another thing that I learned from him, which I found rather remarkable. When I was a kid, my grandfather made me bows and arrows and he used to make the bows out of ironwood. And one of the problems with ironwood bows is that they follow the string. What that means is that although the wood is very strong, over time with the tension of the string, the bow will take the bend because of the string, so you don't get the amount of shooting power. I had a discussion with George about bows and arrows. Of course, George told me that bows and arrows were very important when he was growing up. And the reason was the ammunition was expensive. So all the young boys in the Bear Island area were taught how to shoot with bows and arrows for doing partridge, doing grouse, and rabbits.

So I said to George, "What about this?" And he said, "Here, we make bows out of dry cedar," and he repeatedly told me "dry cedar." And I thought, *Hmmmm, cedar is one of the weakest woods I know of. Why would you make a bow out of dry cedar?* And it's because cedar has that spring back. It has a memory and it'll come back into position. After listening to that, I was motivated enough to get some dry cedar and make some bows to find out exactly how those things work. So I actually made a couple of bows. I can take you downstairs and show the bows that I made. And, sure enough, everything he said was absolutely bang on. Dry cedar is an amazing wood for a bow. You just have to make the arms on the bow thicker. And he showed me how he held the string and all that stuff.

And he went on to talk about the various forms of the arrow. The one they were particularly interested in in Temagami was the blunt-ended arrow, because it was used for knocking down the birds and stunning the rabbits.

And he said, "We also use shell casings in place of the knob. We'd stick that on the end of the arrow to create the necessary weight."

And from there he went on to talk about pole snaring, putting a snare on the end of a twelve-foot pole. This is not ruffed grouse, it was spruce grouse, the dumb ones that sit on the branches on the trees. He could take the pole ... I knew of this of course, I'd read Provencher. But what Provencher fails to tell you in his description is how you actually snared the bird. There's a way it has to be done. You walk around dragging this pole around with a snare on it. But when you're going to snare the bird, you just don't move it over like that and onto the bird. You have to take the pole and point it directly at the bird and gradually work it up that way, point it straight at the bird. And then very carefully loop it around the bird's neck and then snap the noose. I didn't know that at the time. He took the time to explain how that was done.

I was learning on all sorts of levels. With George, I was learning about land use. I found some interesting things that I'd never thought about. Like this exploitation and fallow business that I described earlier, where one half [of a trapline] would be worked and the other half would be let go for a few years. Also in the names. *Shesh-kandika* and *osh-kandika* [new versus established — different types of forest] and all that sort of stuff. They had well-developed concepts of green growth versus old growth, what you find in old growth, where you set up your camp with respect to new growth and old growth, trying to get both. You could trap the martins in the old growth, if that's what you're going for. Or, if you're going for rabbits, you're obviously in the other stuff, or you're after moose or whatever, the vegetative cover was always an important topic of these people. They have all sorts of words to describe forest conditions. All sorts of words to describe snow conditions, travel conditions.

Beyond family and community politics, there were also the complications of working at the Frost Centre, being a provincial employee, and doing research that effectively required Craig to testify against his employer, the Ontario government. Craig had asked his bosses, when he knew a subpoena would soon be served, if this would be okay. He got the all-clear from higher up but, as he sat through seven days' testimony, particularly Tim McCabe's humiliating cross-examination, he must have wondered if it would have been better for him had his OMNR supervisors embargoed his testimony as a conflict of interest. It was only when he started thinking about how he might actually publish this map he'd been working on for all this time that the political consequences of being a government employee and testifying against his employer came to light.

If the trial crystallized anything in Craig's mind, it was that his research and his map had value, particularly for the Teme-Augama Anishnabai. He also realized that this body of work he had been amassing for more than a decade was like nothing else anyone had ever seen, even the academics, and its value extended far beyond Bear Island.

Proof of that came soon after the trial, when Professor Bruce Hodgins, a historian from Trent University and the long-time director of Camp Wanapitei on Ferguson Bay on the northeast reach of Temeeaygaming, asked Craig if he might encapsulate his research in a chapter for a new book Hodgins was working on. This would be a remarkable collection of essays by research colleagues and interdisciplinary friends in the orbit of the Frost Centre for Canadian Studies at Trent University, including Margaret Hobbs, who co-edited the book; John Jennings; Shelagh D. Grant; Gwyneth Hoyle; Jamie Benidickson; Jean Murray Cole; John Marsh; John Wadland; and Hodgins himself. These were ably augmented by two colleagues from Queen's University: William C. James and C.E.S. Franks. The two outliers in this collection were chapters by University of Toronto physics professor, bibliophile, and inveterate wilderness traveller George J. Luste, and a piece by Craig Macdonald, the OMNR recreational specialist from the Leslie M. Frost Centre in Dorset. Craig's contribution was called "The Nastawgan: Traditional Routes of Travel in the Temagami District." Evidence of the continuing rippling impact of Craig's work in the university community was the fact that Hodgins and Hobbs picked up the Anishinaabemowin term that Craig had brought into this conversation and entitled the whole collection *Nastawgan: The Canadian North by Canoe and Snowshoe.*

It was a substantial challenge for Craig to marshal his research into a couple of thousand words that would satisfy him, as well as editors Hodgins and Hobbs. While by now his snowshoe talk was well honed and, since the trial, he was getting better at oral overviews of his work, these were interactive expressions of his research, and Craig could wander as he got a feel for what his audiences were interested in. A master's degree in science, he learned, was not the greatest training in disciplined discursive writing. Nevertheless, with the help of the editors of *Nastawgan* he persisted, and his chapter stands to this day as the first and only published window into his work. It begins:

> *Nastawgan* is a word still in common usage amongst the older, native-speaking Indians of Northeastern Ontario. *Nastawgan* are the ways or the routes for travel through the land of the Temagami district in Northeastern Ontario, just south of the height-of-land.
>
> Before the advent of roads and railways, waterways provided the principal routes for travel and communication over much of the shield country of Northeastern Ontario, including the Temagami and Lady Evelyn watersheds. It was much easier to travel on the waterways than to traverse the rugged, rocky and densely forested terrain. Waterways were used not only in the summer for canoe travel but also in the winter for travel by snowshoe and toboggan. In many instances, the winter routes of travel varied little from those used in the summer.
>
> A trail network based on waterways provided a wealth of natural campsite locations as well as access to fisheries which furnished a major portion of the summer diet for early residents. Unlike land based trails, full exposure to biting insects was limited to campsites and portages. *Onigum* or canoe portage trails were maintained to by-pass

> unnavigable portions of the route: rapids, falls, heights-of-land between waterways and floatwood jams known to fur trade canoemen as "embarrasses."
>
> *Bon-ka-nah* or special winter trails over land were equally important. In winter, the chief obstacles to snow-shoe travel were and still are open water and unsafe ice, found where moving water resists the formation of ice. Sometimes the Algonquin Indians simply extended the trail at the ends of portages in order to reach calmer water with sufficiently thick ice for safe travel. Rivers with strong current extending continuously over long distances posed a greater challenge, especially where there was no space for travel along the shorelines. In such cases they made longer *bon-ka-nah. Bon-ka-nah* were also constructed specifically as shortcuts to reduce the length of winter travel routes.[3]

And on the chapter goes, detailing how *nastawgan* were marked and maintained along with notes on snowshoes, canoe sleds, trail building, how seepage and groundwater affect trails, and where the main intersections were located in the Temagami trail system. However, it is only in the author's note at the end of the chapter that readers get any idea of the breadth of Craig's work and how this contribution to the book differs from all the others in the extent to which it is based on primary ethnographic research.

> Note by the author: The material for this article has been derived as part of a detailed on-going study covering Northwestern Quebec and Northeastern Ontario. Sources include archival map collections, survey records, and especially personal field inspections. Particularly useful were interviews with many of the elder of the Temeaugama Anishinabay on Bear Island and elders of adjacent Indian bands, some of who are now deceased. This invaluable resource has provided deep insight largely unobtainable from the written record.[4]

It is ironic and strangely portentous that a fan of Craig's work in Germany, Erhard Kraus, republished his Nastawgan chapter in an online blog for German wilderness enthusiasts in the early 2000s. A typographic slip in the transcription of the original text changed "detailed" in "detailed on-going study covering Northwestern Quebec and Northeastern Ontario" to "derailed."[5] The fact that there is no mention at all of Craig's map in his book chapter makes "derailed" a more apt modifier for what was going on with his research at that time, pointing to difficulties he was having in knowing how to make the substance of his research available to a larger public audience.

Craig had been coached by cartographers inside the OMNR, including Jack Pound, as he sought to make a permanent record of his research. In 1985, when *Nastawgan* was published, he was still working very much on his own, in isolation in the basement of their little staff house at the Frost Centre. Because of the trial, his bosses at the ministry certainly knew about his work. There may even have been pressure from Ontario justice officials not to encourage or promote Craig's work because of how it was complicating the province's case, which was now heading to the Court of Appeal for Ontario en route to the Supreme Court of Canada.

But that all changed in 1985 when an OMNR colleague in Toronto, Mike Smart, invited Craig to make a presentation about his map to the Ontario Geographic Names Board. As part of the Office of the Surveyor General, Mike Smart served as executive secretary of the Geographic Names Board and could, in this role, extend such an invitation. As an insider with the OMNR's Cartographic Division, Smart also knew that as part of regular procedures the Geographic Names Board was conducting a periodic revision of geographic names in the Temagami district.

So interested was the surveyor general in Craig's work that the 53rd Meeting of the Ontario Geographic Names Board was convened at the Frost Centre. The minutes of the meeting, dated Monday, August 26, and Tuesday, August 27, 1985, show the following in attendance: I.K. Ganton, chair and geographer; B. (Basil) H. Johnston, vice chair and Anishinaabe ethnographer from the Royal Ontario Museum; G. Gauthier, Franco-Ontarian educator; J.H. Warkentin, legendary York University geographer; and L.M. Sebert, topographic engineer, along with Surveyor General S.B. Panting, and the executive secretary and Craig's champion, M.B. (Mike) Smart. There was no doubt that the members of the Geographic Names Board were astonished by what Craig had to say. The minutes on Agenda Item #8, "Role of Oral History in Research Dealing with the Mapping of Traditional Routes and Geographical Names of the Ojibway and Cree (C. Macdonald)," read as follows:

> With the aid of a remarkable map depicting one of his areas of special and intensive in-depth research, Macdonald described how he began his research on the Cree and Ojibway (including the Montagnais of Quebec) with respect to their oral histories, traditional routes of travel (winter & summer) and toponymy. Macdonald's map (of which there are no copies) records the traditional native routes of travel in that part of Ontario bounded by Sudbury in the SW, Gowganda in the NW, New Liskeard in the NE and North Bay in the SE. All information is mapped at a scale of 2" to the mile.
>
> Macdonald explained that the routes depicted included the precise location of more than 75 snowshoe trails and 1000 portages. Native names for the lakes, hills and related geographical features are shown as used and known to the native population. He went on to explain that he had spent an exhaustive 19 years compiling this information. His research involved the examination of all known cartographic material produced of the area in question. Sources included unpublished traders' maps of the period when the Hudson's Bay Company dominated the north; military and non-military exploration record, railway and tourist maps; official government documentation housed in the Public Archives (PAC) to do with geological and topographical surveys, including material in the Provincial Archives and MNR and numerous private collections.

Macdonald pointed out that verification of the location to his data on portages was personally undertaken through nearly 1000 miles of travel by canoe, including numerous investigations by motor boat, snowshoe and snowmobile.

Virtually every available native elder resident, in or formerly a resident of the study area was interviewed. Also interviewed were numerous retired government officials, white trappers and prospectors intimately familiar with the area. Macdonald mentioned approximately 100 native informants. All information, particularly the aboriginal toponymy, was subjected to a thorough cross check for accuracy. Data provided by his informants corresponded well with the records of the late 18th and early 19th centuries, indicating, he said, considerable accuracy, having passed orally through generations of native families without loss of form or meaning.

Macdonald reported that his research is completed. Progress on the map itself to-date has involved preparation of a scribe and peel coat for the water features. Further art work is required to incorporate the portages, snowshoe trails, toponymy and text and map title. This, Macdonald said, *would require the services of a professional cartographer.* On completion of this art work the map will be, he explained, "camera ready" and could be taken directly to the printer for publication....

On behalf of the Board, the Chairman thanked Mr. Macdonald for what was generally received as the most impressive presentation, and expressed hope that, on the strength of the Surveyor General's remark that the Branch would explore the prospects of assisting Macdonald in ensuring that the results of his work were safely duplicated and transferred into the appropriate Ministry records for safekeeping for the generations to come, the powers that be would take the necessary steps to guarantee that such a valuable legacy of Ontario history and geography is not lost.[6]

This meeting was a defining moment for Mr. Macdonald and his map. The Ontario Geographic Names Board, which was an arm's-length organization attached to the OMNR, hemmed and hawed about how to support Craig for another four meetings before — at the 58th Meeting, held in Toronto on December 5, 1986 — its members proposed and carried the following motion: "Moved that the executive secretary of the Board take the necessary steps to have Craig Macdonald's map manuscript prepared in cartographic form with a view to having it published in fiscal year 1987–88."

The Cartographic Division at that time was in the beginning stages of the shift from analogue to digital data manipulation and commercial printing but still had a room full of light tables and very competent old-school cartographers, technicians, and artists who saw to the production of other maps the province wanted for land use and resource management. But the Ontario Geographic Names Board also commissioned — an element of the project that was never realized — a separate accompanying booklet or gazetteer containing all of the Anishinaabemowin names on the map and their meanings. What Craig didn't tell them at the time of his initial presentation was that he hoped and intended to trace all of the shorelines and water levels depicted on the map to their original pre-dam conditions.

What Craig gained in production capacity by signing a legal agreement with the Government of Ontario to publish the map he lost in control and independence, which had been hallmarks of his research since the very beginning. This meant that all the names and papers that had been on his light table in the basement at the Frost Centre were now the responsibility of the map-makers at Queen's Park. And while this might have been reason for Doris and the kids to rejoice, Craig, who had fussed over every detail up to that moment, had to relinquish his design sensibilities to a much more collaborative process involving people who actually made maps for a living.

It was decided that a large-format historical map like this — for which there was no precedent, really, in the Canadian cartographic canon[7] — should have marginal art that would illustrate the canoes, snowshoes, people, clothing, and cultural customs associated with Anishinaabe traditional land use. And happily for Craig, whose childhood art training had rendered him a very competent drawer of stick figures, the principal artist

OMNR artist Alejandro Rabazo got totally on board with Craig's zeal to ensure authenticity of every image created to embellish the *Historical Map of Temagami.* Rabazo rendered the faces from an archival black-and-white photograph but used Craig as a model for other body parts, such as hands.

assigned by OMNR to illustrate and supervise the production of his map was Alejandro Rabazo, a superb watercolour painter who had immigrated to Canada from the Extremadura region of Spain in 1960.

It was clear from the outset that this was not going to be just another forest inventory map. The most complex image Craig thought should be included was a remarkable early-twentieth-century black-and-white photograph of three forebears of his informants — Michelle and Mishaynis Katt and John Paul — processing game at a campsite during an autumn hunt.

Soon after being assigned to the project, Alejandro made a short trek from his office at Queen's Park to the Royal Ontario Museum, where he scrutinized the plumages of game birds to get the colours right for the map illustrations. There were also historic tools and rifles in the ROM collection that Craig held for Alejandro to sketch so that he might perfect those details.

Years later, when I spoke with Alejandro, he laughed, saying that if anyone wonders why all three hunters have the same hands as they hold their equipment, it's because they're all Craig's hands!

For his part, Craig provided as much photographic detail as he could for the map-making team so that the veracity of the art would match the lengths to which he had gone to make sure all the other information on the map was as accurate as humanly possible. Alejandro's camp scene also depicted the making of *shkizzee*, a prized boiled meat delicacy made from the nose of a moose. Nothing would do but for Craig to procure a moose head from one of his hunter friends. This he skinned to get the colour and anatomical detail just right. Happily, said Alejandro, he didn't bring it down to Queen's Park in the back of his car. Instead, Craig borrowed Doris's camera to take a whole series of photographs of the skinned moose head to serve as reference material for the map art.

As 1985 turned into 1986, 1987, and beyond, while the map was out of Craig's sight in Queen's Park, it was anything but out of mind. The minutes of Ontario Geographic Names Board Meeting #63, held on October 21, 1988, in Toronto, indicate that the delays in publication were not only internal at the OMNR: Craig was frustrating the process with his compulsive quest for detail about the original water levels and pre-dam shorelines. The minutes record that the Board's confidence in the project remained strong — "[Macdonald's] map ... will be a toponymic benchmark against which future Native names studies in the Province will be referenced" — but that communications with Craig were deteriorating. The OMNR staff member who gave the status report said that "he had since had a meeting with Mr. Macdonald, in order to identify some of the difficulties in getting from his research material, which was a manuscript map, to the final product. He said it was a complex situation in terms of technologies they were using today, such as Cronaflex to Mylar, Mylar to Cronaflex, Photoseparation, Color Separation, Photo Mechanical Color Separation. He said that it was no longer easy to go from the manuscript to the finished product like they used to."

At Meeting #64, convened on November 30, 1988, members of the board expressed "serious personal reservations about the whole map project, its production, publishing and related financial matters," arguing that "the real problem was that effective control over the project was not possible because one did not have control over Macdonald."

The tone of the minutes for subsequent meetings changed quite dramatically. The board was well aware of the land claim trial — at which their OMNR colleague, Mr. Macdonald, had testified for the plaintiffs — and its pending appeal. They were also aware that in May of 1988 the Teme-Augama Anishnabai Tribal Council had decided at its annual assembly to blockade the Red Squirrel Road extension that had been announced earlier that month by Ontario Minister of Natural Resources Vince Kerrio, Craig's boss. In April, one month before the land claim was due to be reviewed, the same Court of Appeal for Ontario ruled on injunctions brought forward by the province ordering Chief Potts and the Teme-Augama Anishnabai to remove their blockades. The same order compelled the province to stop all construction until the outstanding title issue had been addressed by the court of appeal.

On February 27, 1989, the Court of Appeal for Ontario upheld the lower court decision, forcing the Teme-Augama Anishnabai to apply for leave to the Supreme Court of Canada to rule on the case. Not surprisingly perhaps, at Ontario Geographic Names Board Meeting #65, convened on February 17, 1989, Mr. Macdonald's map was essentially put on ice, as was the original motion to help produce the map and release it in 1987–88. Meeting minutes record what happened: "After some discussion about the status of the map (not yet complete) and the possible prejudicial impact with respect to the Bear Island Band land claims dispute, it was decided that the news release was premature."

Two years later, in the spring of 1991, the Supreme Court of Canada heard the final appeal of the Temagami land claim. That summer the court published a decision rejecting the appeal and upholding the decision of the Court of Appeal for Ontario. And six years after the Ontario Geographic Names Board had commissioned the map, for which all the research was done and all the base layers complete, Craig's map had still not seen the light of day. It would not see the light of day for another four years.

Thanks to the map, however, there was a minor Craig-engineered moment of connection and elation on the emotional roller-coaster ride that was the legal process for Gary Potts and his allies on Bear Island. In his zeal for accuracy, Craig wanted Alejandro Rabazo and the team to create for the map a watercolour image of Second Chief Aleck Paul from a black-and-white photograph taken by American anthropologist Frank Speck on a visit to Bear Island in 1913. Craig was well aware of Speck's work, which had been considered often and in detail during the trial, and he knew that as a travelling anthropologist, Frank Speck had collected different artifacts of material culture, including Chief Paul's ceremonial headdress. Wanting the colours of the painting to be accurate, in his quest for perfection, Craig went on a hunt to find the headdress.

Through correspondence with the American Philosophical Society in Philadelphia, Craig learned that the headdress had been repatriated to what was then the Museum of Civilization in Hull, Quebec. Redirecting his inquiry, Craig connected with the curators and informed them that a Teme-Augama Anishnabai delegation would be attending the Supreme Court of Canada to witness their case going before the court. After a protracted series of negotiations, Craig managed to persuade the curators that Chief Potts might like to don that sacred item from the past as a talisman for the future that was being decided in Ottawa.

Reluctantly, two curators brought the headdress to the Ottawa hotel where the delegation was staying. Craig described to me what happened next: "They opened the thing up and plopped the headdress on Gary Potts's head. They did their thing, took the photographs. Whatever. They explained that feathers are very delicate. It had to be in museum-type conditions if it was going to survive. They took it off his head and gently put it away. So, anyways, I got pictures and was able to get the exact stuff. So that's the real McCoy on the map, with the beading pattern and so on."

What may have been lost on the federal curators, for whom the headdress was a valuable piece of Anishinaabe material culture completely separated from any meaningful cultural context, was that the fight Second Chief Aleck Paul had been fighting in 1913 — namely, for recognition from the governments of the day and a place for his people to call their own — was the exact same battle being waged by Chief Potts and the delegation. Although it was Craig who brought to the delegation's attention the location

of the headdress in Ottawa, there was a through-line of tradition that had been kept strong in the place names, practices, and traditions that were still driving the Teme-Augama Anishnabai nearly a century later. There's no doubt that physically connecting Chief Potts to Chief Paul's historic headdress connected them all to words Chief Paul spoke in 1913, which were recorded by Frank Speck:

> When the white people came, they commenced killing all the game. They left nothing on purpose to breed and keep us the supply, because the white man don't care about the animals. They are after the money. After the white man kills all the game in one place, he can take the train and go three hundred miles or more to another and do the same there.
>
> But the Indian cannot do that. He must stay on his own section all the time and support his family on what it produces. If an Indian went to the old country and sold hunting licenses to the old country people for them to hunt on their own land, the white people would not stand for that.
>
> What we Indians want is for the Government to stop the white people killing our game, as they do it for sport and not for support. We Indians do not need to be watched about protecting the game; we must protect the game or

> starve. We can take care of the game just as well as the game warden and better, because we are going to live here all the time.[8]

Everyone who remembered Aleck Paul had great respect for his determination in the troubled years of the early twentieth century. When archaeologist Thor Conway posted Aleck Paul's infamous speech to Frank Speck on his "Temagami Secrets" Facebook page on April 22, 2020, he added, "We need to remember his poetic, indigenous name — Ozheshewakwasino-Winini, 'The Man Who is the Sound of the Wind Rustling in the Trees.'" Craig's one big regret in creating the map was that Second Chief Aleck Paul died before he was able to talk with him.

Chief Paul's concerns about conflicting messages from governments about the shared use of the Teme-Augama's ancestral homelands might well have been on Craig's mind when he learned that down at Queen's Park, within the Office of the Surveyor General, a young new bureaucrat had been assigned to the management of the map project. Ten years into her career as a geographer, Elizabeth Blount[9] had been hired to help with the transition from analogue to digital mapping. She was charged with raising awareness both inside and outside the OMNR of Geographic Information Systems and with building the digital geographic databases used by the emerging digital apps and navigation systems. By accident or design, the *Historical Map of Temagami* became part of her brief.

This was the beginning of the end for the Ontario government's belief in the map project. After the Ontario Geographic Names Board's reservations had been laid down in the meeting minutes, word was that Ms. Blount was not happy with having *anybody's* name on Craig's map, including his. In his one meeting with the new supervisor, Craig mentioned that he had a contract with the government that said otherwise. Ms. Blount's line of thinking persisted, however. She told Alejandro Rabazo and Joe Okorn, the other artist who had worked on the map, that they would not be signing their work, as was customary. But for years Alejandro had been secretly burying his children's names in the shading of other maps he had created, so that only they would know. Now, he and Joe cagily disguised their signatures in the images they painted for the map, conveniently placed somewhere other than a bottom corner, where a viewer might expect to see an attribution.

As the 1980s turned into the 1990s and the map still was not published, Alejandro feared it would *never* be published. When I spoke with him in June 2022, he was eighty-eight and a half years old and still as feisty as ever. He recounted to me a three-page handwritten letter he sent to the minister of natural resources, imploring him to keep the project alive. "Craig Macdonald has been working on this project for twenty-two years," Alejandro wrote the minister, adding that much excellent work had been done by Mike Smart and the OMNR team, and that they were poised to put the map on one of the ministry's big presses to bring it into the light of day after all these years of development. Looking back now, if it weren't for that letter, or perhaps a change of heart on the part of ministry personnel overseeing the project, the *Historical Map of Temagami* might never have been published.

Perhaps by total coincidence or perhaps serendipity, the Supreme Court of Canada delivered its decision in 1991, and the map was finally released in 1993. Happily, with Craig's name on it. However, Alejandro Rabazo's name did not appear on his illustrations — at least, not so that a casual viewer would notice, anyway — but he did prevail with recognizing at least some of the people who had worked on the map in a little window in the bottom right corner of the huge sheet:

> Cartographic production and publication commissioned by the Ontario Geographic Names Board, August, 1985.
>
> Project manager: Michael B. Smart
> Cartographic supervisor: Alejandro Rabazo
> Cartographic technicians: Jeff Bender
> Chris Markow
> Fanny Paz
> Alejandro Rabazo
> Illustrators: Joe Okorn, Alejandro Rabazo
> Photomechanic technician: Barry Rudge

No mention of the Cartographic Division of the Office of the Surveyor General, the Ontario Ministry of Natural Resources, or Elizabeth Blount.

Normally, after a long, involved groundbreaking project like this, even within government, there would be some kind of launch and celebration of the publication. There was nothing of the kind. Not even a press release published late on a Friday afternoon. Of the ten thousand copies printed, several thousand were allocated to Craig (who had to pay $2,000 to cover the cost of the printing), as per his agreement with the ministry, and the rest were kept at Queen's Park. Record copies were sent to Library and Archives Canada; the Smithsonian Institution in Washington, D.C.; and the United States Board on Geographic Names. Sadly, the copies in the government's inventory quietly disappeared. Both Craig and Alejandro feel they may have been shredded or incinerated to make room for new digital mapping equipment in the OMNR Cartographic Division offices, or maybe just to get Mr. Macdonald's map out of the office once and for all. Either way, the OMNR might well have been glad to be done with the map contract and to have Craig Kennedy Macdonald on the books solely as an OMNR employee lost in the wilds of Algonquin Park.

Travelling with Craig III — In the Land of the Montagnais

TO UNDERSTAND CRAIG MACDONALD IS TO GO BACK TO THE 1980S and appreciate that his interest in Indigenous travel routes and technologies went far beyond the creation of a map of one relatively small portion of his research area. There was no question that he was rocked by the political grindings of the trial and the frustrating delays in the production of his map by the OMNR, but in spite of all that Craig remained singularly focused on continuing his research.

This included an ongoing interview schedule but also further trips to try out some of the things he was learning from his informants and from his library and archive explorations. As when he was fixated on the massive pine tree on a distant ridge when he was a young staff member at Camp Kandalore, now he was focused on the goal of a comprehensive understanding of traditional human striving through the full extent of the boreal biome from coast to coast to coast.

And while the interviewing and reading fed his intellectual curiosities, time on the trail was somehow more satisfying for Craig, perhaps because it took him away from the complexities of work and family and out of range of the bickering governments and the courts. Simply put, there was nothing more satisfying for Craig Macdonald than waking up in a tent, preferably in the winter.

Reflecting on his dad's essential nature, his son, Colin, had this to say:

"I often wonder if that man who went to Kandalore went on a canoe trip that has never ended.... I think Kandalore had a huge impact on him. And I don't think he ever let that go. He never turned the page. He never left that canoe trip. His mind was always on that canoe trip. Through fatherhood, he had his wife and his children. When he goes to bed he's still thinking about his canoe trip. And his love for winter camping has to do with having that canoe trip go on right through the winter."[1]

After his experience of walking the wind-packed snows of James Bay in 1982, Craig got it in his mind that in 1983 he needed to follow his hero Paul Provencher into the deep snows of Innu country on the north shore of the Gulf of St. Lawrence. I received a classic Craig letter in the mail in December 1982. The original, as always, was written in pencil on foolscap and was probably copied in the main office at the Frost Centre. It read as follows:

> Frost Centre
> Dorset, Ontario
> P0A 1E0
> December 12, 1982
>
> Dear Participant,
> I am now assembling supplies and equipment for our Laurentide Park Snowshoe Expedition. To this end, *I will require as soon as possible, a $200 deposit.* Any monies not utilized will be returned after completion of the trip.
>
> If for any reason you are forced to drop out of this trip, please notify me as soon as possible so there is adequate time to recruit a suitable replacement.
>
> Extreme elevation gains and deep snow make the hauling of every additional pound significant. Therefore, please adhere to the enclosed personal equipment list as closely as possible.
>
> We will be working together as a team, taking turns breaking trail and hauling the various loads for the duration of the trip. Because of the danger and technical

supervision required, especially on the river sections, we intend to travel as a "tight-knit" unit[2] to maintain the best communication possible. However, it may be necessary early in the trip to send two people ahead of the main party each morning to "freeze-in" the track. Hopefully there will be sufficient time later in the trip for people to take a few exploratory side trips without the burden of equipment from one of the encampments.

I am enclosing the trip itinerary. Please note that you should carry enough money for at least 10 meals on the road as well as up to 3 nights of motel accommodation.

This should be a very rewarding and challenging trip. There will be opportunity to learn a whole new range of technical skills as well as to experience first hand what winter travel is like in the land of the Montagnais.

Yours sincerely,
Craig Macdonald

Personal Equipment List

1 sleeping bag (good to 0°F)
1 closed cell foam pad (eg. ensolite) NB. # keep width to 23" — if necessary thin pad down with a hacksaw or coping saw — disregard the above if using an air-foam pad
1 waterproof ground sheet *or* duffle to protect sleeping bag in transit
1 bic lighter *and or* small waterproof package of matches to be carried on the body at all times
1 jack knife or pen knife or small clasp knife etc.
1 pair of rubberized waterproof footwear (4 buckle boots, or rubber-bottoms leather-tops or snowmobile boots etc.)
2 pairs of removable felt boot liners

1 pair of moccasins or mukluks for long distance snowshoeing in dry conditions. ← try to get a very light weight pair
6 pair of woolen socks
3 pairs of long underwear
3 pairs of short underwear
1 pair of heavy wool pants or warm equivalent
1 pair cooler pants (possibly jeans or any other relatively windproof material)
1 pair of moderately large snowshoes with harness for this trip (avoid excessively long tailed designs as these will be very bad on the hills)
2 pairs of large mittens (one pair preferably with removable liners if your hands sweat)
1 really warm hat or touque (2 hats if made of light material)
1 large windproof scarf or possibly balaclava to protect face.
1 heavy wool sweater
1 winter coat that can be opened up at the bottom for ventilation yet closed at the neck. #N.B. *even in severely cold temperatures traveling through the forest sections we will be boiling hot especially on the hills.*
1 pair of polarized sunglasses (*clear days present a severe snow-blindness hazard*)
1 small flash light
1 thermos (unbreakable plastic inners better than glass as sleigh rollovers will be frequent)
Plus Duffels to contain the clothing items above.

Land of the Montagnais
Laurentide Park Snowshoe Trip '83
Itinerary

Thurs Feb. 17
Everybody gets into my suburban truck and we drive to Quebec City where we pick up Jake Fallis at the airport that night. — spend the night in motel in the general area.
Friday Feb 18
— meet park superintendent and start trip in early afternoon
Finish Trip — possibly very late Thursday Feb 25 or more likely either Fri. Feb. 26 or early Sat. Feb. 27.
Sat. Feb 27 + Sun. Feb. 28 — drive home (at least one motel en route).

The insanity of what Craig had cooked up this time was delicious! As per instructions, five of us — Bill King, a vascular surgeon from North York; Peter Attfield, an outdoor education technician with the Toronto and Region Conservation Authority; Mark Scriver, an outdoor gear sales representative; Craig; and I — gathered at my house north of Kingston, knowing we would catch up with the sixth and final member of the expedition, high school principal John (Jake) Fallis, the following day in Quebec City. Since doing the initial planning for the trip, Craig had continued his research, and he did his best on the drive to emphasize how potentially difficult the terrain would be. Like the rest of the group, I had a copy of the 1:50,000 scale topographic maps covering our route which, to be fair, were more brown with densely packed contour lines than green as they had been for the James Bay trek. But we had *no idea* when Craig wrote in his letter that we would encounter "extreme elevation gains and deep snows" how absolutely indelible these aspects of this journey would be. No idea.

As for the deep snow, we were routinely breaking trail through snow so deep and soft that the snowshoe track would be more than knee-deep before there was sufficient compression to support the weight of a person.

This was one of the things Craig wanted to experience — the deep snows of Montagnais country as described by Paul Provencher. A note from my journal from that trip details the totally surprising depth of snow and the challenging conditions for hauling at the bottom of this fifteen-hundred-foot cleft in the boreal Precambrian Shield:

> Because the valley here is so steep, the walking is terrible and the hauling is obscene. Without doubt, the craziest feeling I had today was while breaking trail *over* 5–8' spruce trees. The snow had drifted in and frozen around these boughs and we were breaking through 18" of powder to walk across the *tips* of the trees! Wow! That was something that was a mind-boggling experience. It wasn't so much that we were walking over good-sized Christmas trees, it was that, from time to time, a hole would emerge and we could see down toward the ground. Sometime as deep as six or seven feet we could see down. The thought of falling into a hole like that was not thrilling.

Besides the prodigious amount of snow that traditional travellers would have to negotiate in this part of the Shield, Craig was intent on exploring how traditional sleds would hold up going through this deeply dissected terrain. And he also wanted to understand — this was partly why he chose to travel through the valley of the Jacques-Cartier River — what it would have been like pulling loads in places where there was never flat ground. "Side hill" was the term Craig used to describe places where the team would be compelled to move not up or down an incline, but laterally across one, with gravity pulling the sleds downward as we tried to pull them across the hill. The solution to that problem, as it turned out, was to get out the shovels. We would dig and pack a narrow track in the deep snow, as they might in the Himalayas or the Andes, to create a flat surface that would accommodate our sleds.

Craig was also interested in learning first-hand how to negotiate steep hills, and in particular how we might get down to the Jacques-Cartier River from the Precambrian plateau fifteen hundred feet above. He had read about ways to foul the runners of the sleds by wrapping them with rope so they would catch the snow and not slide down out of control, but he needed to try out this technique to fully understand the mechanics and challenges of travelling in this country. This story became a feature of the letter I wrote immediately following the trip to my old friend Grace Reid:

> Box 18, RR#2
> Lyndhurst, Ontario
> K0E 1N0
> February 28th, 1983
>
> Dear Grace:
> I've just returned from this year's snowshoe stomp in the hills north of Quebec City and in between sorting out bills and other mail I thought I'd drop you a note to tell you about the adventure.
>
> This was another junket organized by Craig Macdonald, the same guy who expedited the James Bay trip. This time, instead of following the snowshoe route of the Hudson's Bay Company packeteers, we were on the trail of Jesuit fathers who, in the 18th and 19th centuries, made regular journeys between Quebec City … and the Montagnais Indian settlements in the Lac St. Jean environs. The part of this ancient route that we travelled included a trek into, and then down, the valley of the Jacques Cartier River.
>
> Looking at the aerial photographs of the valley before going on the trip in stereo (3 dimensions) it was clear that this was *some* valley. Even though it was formed by glaciers, it was over fifteen hundred feet deep and in the upper reaches only eighty metres wide at the bottom. Picture a knife-like slit in the Precambrian plateau of the Laurentian Shield and you'll have a pretty good idea of what the Jacques Cartier River flows through. The problems posed

by this topography for our toboggan and sled/snowshoe mode of travel were numerous and quite perplexing. We could use old logging roads to get to the edge of the valley, but these ended abruptly at the precipice. Somehow we had to get our stuff down fifteen hundred feet of near-vertical valley wall. And then, when we would get to the bottom, there was still the problem of traversing these steep valley sides to follow the river. You see, the river was dropping so quickly (50' @ half mile) that there was no ice to walk on and there was no flat bank on which to drag our sleds and toboggans (collectively the Indians referred to these snow vehicles as "odawban"), the steep walls of the valley just continued right into the raging water. Some valley indeed!

Well, the six of us set out along the old logging roads and quickly learned that although most of Ontario was snowless, the region north of Quebec was loaded with the stuff! They normally get over three hundred inches of snow in a winter and this year was a dry year, but we still had to break a trail through six or seven feet of the white stuff from time to time. We travelled in single file, five people dragging an odawban and one out in front breaking trail. At half hour intervals (or sooner if the conditions were tough for the trail breaker) we rotated in what Craig called "our workup." The trail breaker went to the back of the line where the going was relatively easy and the rest moved up one notch in the line. This type of travel continued without interruption until our route was impeded by thickets of balsam fir saplings. When these were encountered we had to send two people out early each morning to stay an hour ahead of the group and cut a trail through which to pull the rest of the gear. This advance guard served two important purposes. It saved time and, more importantly, it allowed the trail to harden before the others came through with the heavy loads. With the deep snow, this hardened or "frozen in" trail made quite a difference. Tough as this type of hack and slash travel was, it was nothing compared to the shenanigans which went on when we tried to get down the sheer valley walls.

Using the air photos we chose a route down the valley wall. We began by descending on the frozen rocks and surfaces of a stream, a process which was as close to going through the Khumbu Icefall at the foot of Everest as anything I've ever done — what with all of the ice bridges and great holes in the snow. When the stream got too steep to negotiate, we moved out of its little valley and onto the land. This was so steep that we had to take off our snowshoes immediately and wade through waist deep snow just to stop ourselves from falling downhill. It was really more of an intellectual challenge than anything else to get our odawban safely down this steep hill. Eventually, we settled on a system whereby using sixty foot lengths of rope we literally lowered our laden odawban from tree to tree. This went on for nearly an entire day. At one point Craig fell and badly bruised his shin to the point that he thought he would be unable to continue but fortunately Bill King, the expedition doctor, shoved some pills down his throat and he was back on his feet again.

Down and down and down we went through forests of black spruce so laden with beard moss that we could barely see the sky. As we got lower into the valley the vegetation began to change to yellow birch and pin cherry and for the first time we got a glimpse of what we were descending into. Through the snow which began to fall we gradually realized that what we were seeing through the branches of the deciduous trees was not the sky but the opposite wall of the valley — rising so sharply and closely that it seemed one could hit it with a long pole! Soon after seeing the opposite wall (we were unable to see the bottom, though we could hear the rushing water) we heard a helicopter flying down the valley — and it was well below where we were standing! Arriving intact at the river's edge we looked left and we looked right on the valley wall to see sheer rock faces on both sides of our descent corridor. More by good luck than good management we had narrowly missed descending to the edge of impassable cliffs. Pheew! Some valley!

It snowed for the five days during which we made our way down the edge of the river. Even the weather, though, could not conceal the splendor of the valley. At each corner new sights would emerge — rock faces, treed cliffs, far distant blue hills, ice rimed trees and sparkling winter rapids with glazed rocks — sights that won't soon fade in my mind's eye.

The odd thing about this region was that outside the valley, the land was ravaged by indiscriminate and badly managed logging and great swaths of treeless land laid bare to allow high tension power transmission lines from Churchill Falls and Saguenay power plants to take northern electricity to southern markets. The area we were in was in a Quebec land preserve called Jacques Cartier Park, but it didn't seem right to have these scarred lands included in the same jurisdiction as the magnificent Jacques Cartier valley. It almost looked like the Quebec government gave the land park status when it was of no further use to them for resource extraction. Odd!...

Salut (as they say *en français*),
Jim

As on other winter trips with Craig, after hard physical days on the trail we would hunker into the heated tent for a meal of Doris's hearty home cooking by candlelight and then collapse into our sleeping bags in the dark, where we'd chat before dropping off one by one, depending on the exertions of the day. It was in this context, with the tent lit only by the cherry-red glow of the stove radiating up from the frost trough, that no matter what the topic of conversation, Craig would extemporize a tale or two from his research travels to link where we were and what we were doing to the days of old, when hardy travellers with the same kit would create the only winter connection and communication lines from place to place, from community to community.

This trip, we learned that Craig's hero, Paul Provencher, "the last of the *coureurs de bois*," was born in Trois-Rivières in 1902, and that, after studying at the seminary there, he attended Laval University in Quebec City to study engineering. Then he began a career with various big lumber companies, carrying out forest inventories throughout the North Shore. It all sounded very familiar: a guy with a fancy university degree who loved the woods — the piquant aroma of smoke from a spruce fire, the lopsided millinery of new snow on rugged scapes — almost more than life itself.

Like Craig, Provencher sought Innu dwellers of the land to learn from. He interviewed Montagnais and Naskapi Elders. He plotted their routes on topographic maps. He learned the techniques and technology of bow-hunting, fire lighting, shelter building, trapping, and so much more. And he took photographs (Craig never really took up that practice in his research), creating a legacy and a body of work, invaluable documentation of Innu life in the 1930s, that Craig truly admired. And just to make sure we weren't dozing off thinking that Craig was making all this up, he would trot out beguilingly specific details like the fact that two of Provencher's key informants and travel companions were Joe "Uapistan" Savard and Ti-Basse Saint-Onge.

These bedtime soliloquies wandered freely, as did Craig's mind. One minute we'd be hearing about James Evans, a Methodist minister, missionary, linguist, and teacher who developed the Cree syllabic alphabet in Rossville, Manitoba. The next, he'd be parsing Anishinaabemowin terms for things that were part of this silent beautiful wilderness world we were in. Stove — *keejab kisigan*. Sliding device — *odawban*. Toboggan — *nabug-odawban*. Then a lesson about not hanging the toboggans at night with their bottoms facing south because this might offend Zhaawani-Noodin, the south wind (the equivalent, he told us, of mooning the Queen), which would bring warm winds and create conditions that would cause the bottoms of our toboggans to collect ice crystals, making them nearly impossible to pull. If that happened, Craig explained, it would be no one's fault but our own, because we hung the toboggans incorrectly.

In the morning we'd put our toboggans on the ground and turn them over so that we could use a pot handle or the axe to scrape them clean of ice. With its bottom side up, he explained, the curve of the *nabug-odawban* looked to his Anishinaabe friends like the sinuous neck of a swan, *wabasee.* But in the parlance of the good people of Bear Island and beyond, *wabasee nabug-odaban*, literally translated to "swan toboggan," actually meant "Oh, shit, I've made a silly mistake." From that day on, when something untoward happened, we'd utter — or rather stutter — "*Wab-a-see na-bug o-daw-ban,*" a very satisfying expletive.

Anthropologist Frank Speck, we learned — the same Frank Speck who was in Temagami in the summer of 1913 — was on the North Shore in 1912, collecting "specimens" for the distinguished linguist and anthropologist Edward Sapir and the newly opened Victoria Memorial Museum in Ottawa (now the Museum of Nature). Speck got involved as a mediator between the Innu and the federal government on the issue of fishing rights and territories in the Moisie River valley, just north of where we were. Despite being a guy who seemed to prefer action to reflection, movement to stasis, Craig had covered great swaths of intellectual ground hunting for detail about travel routes, methods, and technologies. But the thing that never ceased to amaze us was his ability to recall the minutiae of material he'd read or heard months or often years before!

In 1984, Craig stuck closer to home, taking trips around the Frost Centre to test out canoe sleds he was recreating from his research. These were *odawban* that could be used mostly in the spring, when the lake ice was melting. In the event that it broke through the ice, a couple of knots could be untied allowing the sled to float up beside the canoe. By pulling four shims, the sled folded like a parallelogram and stored in the canoe, allowing the traveller to continue with a paddle instead of a tumpline.

In the autumn of that year, another classic Craig invitation arrived in the mail. This time he was proposing a journey into the country north of Lake

Superior, terrain touching on the lands described to him by Joe Bananish in the infamous eight-hour Longlac laundromat debacle.

> Feb. 2–Feb 10 Lake Superior Park Snowshoe Expedition
> Organizer: Craig Macdonald
> Book as soon as possible
> I am planning a snowshoe expedition crossing a large section of Lake Superior Provincial Park using traditional techniques including Indian style tent stoves and odawbans. We will travel north on the Algona Central Railway from Sault Ste Marie through the Agawa Canyon to Mile 136 ¼ at Sand L. From here we will snowshoe down the Sand River then westward through several watersheds to our destiny on the Trans-Canada Hwy #17. To ensure high quality outfitting, all food + equipment will be supplied by myself. Participants will only be responsible for their personal clothing, foam pad + snowshoes. All costs will be shared. On the evening of Friday Feb 1 we will rendezvous at my home to complete packing + loading. Seven days will be spent on the trail.
>
> This trip promises great scenery, plenty of snow (highest snowfall area of Ontario) and excellent opportunities for ice fishing in the more inaccessible lakes. Lake Superior Provincial Park is ideally suited to recreational snowshoe camping and on our route we will be a pioneering attempt to explore some of this potential. With a small commitment to pre-conditioning on snowshoes, this adventure is well within the physical capability of most W.C.A. [Wilderness Canoe Association] members. Extensive winter camping skills are not required as instruction covering the necessary skills will be given en route. A great trip!

Not long after this invitation arrived — it might have been just after Christmas — Craig called to say that Doris's health had taken a turn for the worse and he felt he had better stay at her side in February. Reluctant to cancel the trip because a number of people had already committed to it and had rearranged their schedules to join him in February, he offered his tent, stove, and *odawban* if I would step in and help make the trip happen.

Craig went on to explain that, unbeknownst to the rest of us, Doris had struggled mightily with migraines and a mysterious bowel ailment that seemed to get worse with the birth of each of their children. Nancy, Janet, and Colin were eight, five, and four years old that January, and their care, feeding, and entertainment were probably more than a full-time job in themselves. Until she was able to get back to herself again health-wise, there was no way she could prepare the food for this trip. Oh, and Craig thought he might be able to help around the house while Doris was under the weather.

I assured Craig that staying home was exactly the right thing for him to do and that we would do our best to carry on without him. What was different about this trip, however, compared to previous ones, was that his work for the OMNR at the Frost Centre had expanded to this kind of research and exploration. I told him that we would do our best to gather whatever information he needed to make his assessment about the recreational potential for winter travel in Lake Superior Provincial Park. He said that, as a favour to him, one of the OMNR's real old-timers, who had grown up on Lake Temagami, a colourful character called Tom Linklater, had agreed to come along to help make that assessment. Craig wasn't sure how he'd fit in with the rest of us, but he hoped it would be okay.

The trip went very differently than planned or expected — the Sand River was impassable because of open water and slush — but thanks to information on a map in the park office of all the logging roads in Lake Superior Provincial Park (which we were able to access and transcribe onto our maps because Tom Linklater was a great friend of the superintendent), we were able to choose an alternative route off the river and still have a great trek from the Superior highlands down to Gargantua on Big Lake Gitchi Gami itself. And, as per our commitment to relay what we'd found to Craig, I wrote to him on Valentine's Day, 1985, just days after getting home:

Dear Craig:
In spite of the hearts and flowers that we're all supposed to be thinking about today, I find myself still thinking about the Lake Superior Provincial Park snowshoe stomp. It really was a nice break in the winter term. To say the least, the atmosphere in these hallowed halls of higher learning can be a tad confining and any chance to get out into the fresh air is welcome. The odawban trek across the park was just what the doctor ordered for clearing out the pipes (and a wee bit of the spare tire that has been growing around my middle).

Your fears about Tom Linklater not fitting into the group because of his age and background were totally unfounded. It was such an eclectic assortment of individuals that gathered in Sudbury for a last minute packing on Friday, February 1st that I'm sure anybody would have fit in somehow. Tom was a solid member of the group in every way. He pulled his weight and did his share, and more, around the campsites and contributed in a constructive way to the decision making process. On top of all that, as I told you on the phone, he was a barrel of laughs. Never before have I come across anyone who could slay a group with one-liners at any time of the day or night. You'd have loved his convoluted sense of humour. For example, in watching one of us fumble with the ropes, trying to untie a lashing with mitts on, he would say, "The only thing you can do with mitts on is shit your pants!" And then, when he tasted this year's rendition of "Bumbo" he said, "Gees, that tastes just like Corby's whiskey and like love in a canoe ... f@#$%^ near water!"[3]

Under ideal conditions, the route you chose would have been fine. But as I told you on the phone, we ran into bad slush problems and had to get up on the logging roads where the snow was deep but the going was consistent. Even at that, it was touch and go for a while whether we'd have to turn around at some point and go down on the train. I think it was the thought of having to eat a large helping of crow from the guys working on the train that kept us pushing through to the highway. Apparently, they'd let a couple of guys off the week before in the vicinity of the Kawagama Lake Lodge, and despite good intentions to camp for a week, the pair was back on the train the next day — having limped into the Lodge in the middle of the night.

The snow depths we were into varied as the snow aged. At the beginning of the trip we were wading through three to four feet of relatively new snow. In these conditions, the only one of us with sufficiently large snowshoes was Tom. We took advantage of this by getting him to go ahead in the evenings, while the rest of us set up camp, to pack a trail that would freeze in by morning. Even with this effort, some of the rest of us had trouble with break-through. As time progressed, however, the snow settled and eventually we were cruising along with only moderate effort. I think that what we learned from the difficulties in the deep snow was that in future we should make an effort to ensure that everyone has big snowshoes and pretty much the same per area force on the snow (i.e. lighter people having smaller snowshoes).

We had no trouble with wood, either for poles or fuel. We did cut some live material for poles but this was usually done in areas where no one would ever go in the summer. We might have used some of the coniferous poles that have been killed by the various attacks of the spruce budworm, but these poles generally had been dead long enough to make them too punky to hold much weight. Fuel wood was always available in some form or other. Early in the trip, some of the new people cut lots of shitty balsam and black spruce which, they found out, made terrible firewood. With some of the stuff that was cut, we would get only about an hour and a half between stokes during the night. As people learned what to cut, we eventually ended up with lesser amounts of better quality fuel wood. And by the end, we were cutting only a small amount of seasoned hardwood that would give us about five to six hours between stokes. One night in the middle of the trip, I cut down a seasoned and pitch-rich tamarack which just about burned the fire trays out of the stove, but other than that the temperatures in the tent was always clement.

I told you about the problem we had with my toboggan, which broke first in half (we fixed it with machine screws from the repair kit) and then the hood came adrift. This we left off (we're keeping this as the makings of the C.K. Macdonald memorial odawban trophy) and to make the device haulable, we simply tied up the side lines in such a way as to make the leading end of the toboggan rise up over the snow. I know the toboggan problems were probably the fault of the way the device was stored, but other problems we had were, I think, largely the fault of one of the expedition members [who I'll call Four-by-Five[4]], who had the gentility of a two-ton bull in a china shop. Good old Four-by-Five managed to break a number of things, including a swede saw, through his brute force and lack of forethought about who else might benefit from the well-being of various bits of equipment. He tended to use things as if he was the last and only one who would ever need to do so. As you would, I gave him some gentle hints about streamlining his techniques but I'm afraid Four-by-Five is not one for subtleties. Most of the "gentle" suggestions from the group went clean over his balding pate. Other than this and the fact that poor old Four-by-Five did S.F.A. around the campsite, the group meshed exceptionally well.

The MNR people in the park office, notably Ray and Pete, were most helpful. This is probably due to your connections and Tom's clout. In any case, it was great to have helicopter support in the event of an emergency and to have the use of the staff house on the last night. Also, their cooperation in parking and shuttling the vehicle was a real help to the smooth running of logistics....

In all, the Lake Superior Provincial Park is a great region for odawban style treks. In fact, I think given the fact that the forest has been so ravaged by loggers and the budworm, the winter may be *the* time to have a look at the park interior. For sure, snowshoeing should be encouraged. The routes could be short and thereby easy to manage and they'd go through territory that would really give the feeling of history and of wilderness. The clincher on the area for this kind of travel is train access. I don't think I can emphasize enough how important it is to the feeling of isolation, that you begin a back country expedition from an airplane or a train. I'll never forget what it felt like to jump through the steam from a warm ACR boxcar stopped at Mile 136 ¼ and to land in waist deep snow. Beginning an expedition like this was really exhilarating.

Tom and I talked about the possibility of putting together a plan for the development of snowshoe routes and support services in the park. If there could be some money made available from the Ministry or somewhere, I'd be really interested in doing some planning with you and/or Tom to explore the possibility of making our route and expedition more accessible to others who might be interested.

You would have enjoyed the stomp and it was too bad that you had to send us along without the C.K. Macdonald expertise. I hope this letter gives you some idea of what things were like along the way. As I told

> you on the phone, the six of us are getting together for a reunion evening on Saturday, March 16th. We'd very much like to have you and Doris along, if it's possible and if she's up to it.
>
> I'll be up sometime soon with your sleds etc. In the meantime, please give my best to Doris and the kids.
>
> Regards,
>
> Jim

Doris's health and his work kept Craig closer to home again the following winter. But as if the yen to follow his bliss to the north powered his need to venture, after two years of research in the immediate vicinity of the Frost Centre, Craig was planning the mother of all (wacky) research junkets, this time to Nouveau-Québec, where he could ask similar questions of his technologies and techniques that he'd asked on the taiga journey from Fort Albany to Moosonee five years previously. What would it be like to travel, as the Inuit and Cree from Great Whale River had done for years, from Clearwater Lake to the Richmond Gulf (now Lake Tasiujaq) on the Eastmain shore of Hudson Bay? Would there be sufficient wood for fuel and tent poles? Would the terrain support a snowshoe-based expedition at this latitude?

Working backwards, the answers to these questions, discovered when Craig and three other hardy souls[5] made the trek in the winter of 1987 were "yes" and "yes" but "painful as hell and not without nearly dying on a last-ditch two-day sprint to get to safety because the people you'd planned to meet at the end of your trek to drive you by snowmachine to the Inuit community of Imiujaq didn't show up."[6]

When I interviewed Doris for this book, I said, "You've got three kids in elementary school. You've got migraines. You've got ulcerated colitis. How did you do it?" This was her reply:

> And I have a husband who's going off on winter trips. I'm doing all the cooking for it. And then, preparing for that James Bay trip, we learned our house was insulated with urea-formaldehyde foam that had to be taken out. So they

had to take all the siding off, pull out the formaldehyde insulation, and then put the siding back on again. He's up on James Bay. I have the kids. They needed to have afternoon sleeps. And finally I went out to the guys and said, "Could you possibly work now and take your lunch hour later so I can get this child down for a sleep?" They were good. They did it for me. But it was rough. But we froze. It was wintertime. We used up a whole tank of oil in the three weeks you were gone.

I'll get to the point. I sent him off saying, "I don't care what happens to you, Craig. Just make sure you kill yourself if something happens because I don't want to have to look after you, an invalid."

I remember one story. I don't remember where he was. Away up north someplace. One of the guys who was with him had told his wife that he would be phoning her on a certain day at a certain time when they would be off the trip. So she phones me because there is no word. And I said, "Well, no. Anything could happen. It might be two days before we hear from them. They could hit winter storms. They could be stranded. They couldn't get moving." I told her I sent extra food with them, so that if they get stranded they've got food. And then she proceeded to tell me, "Well, I just got back from Florida." And I said, "Well, I'd like to have been sunning in Florida too but instead I'm here running a household and looking after three kids. If I don't hear from them in a couple of days, I'll call somebody to find out where they are. I wouldn't worry about this."

I remember George Hamilton [Craig's boss at the Frost Centre] calling me one day and saying that people in an MNR plane had seen you guys somewhere on the frozen Magnetawan River. And I said, "Oh fine. Good. I don't really care. I don't have time." So, yeah, that's the way I lived with it. I just couldn't worry about him.

> The only time I think I really started to worry was the time he was walking across Richmond Gulf. And then when he phoned me and told me they'd gotten lost and he and another guy had gone off on their own and, you know, they were stranded out there in the middle of nowhere. And when he phoned, I thought *Oooooohhhh, that's kind of serious.* But I just couldn't think about it. I just had too many other things on my mind. You do this, Craig. Go. I'm supporting you on that, but I can't worry about you too.[7]

Doris did have a limit, however. Once, out of frustration after Craig had been more absent than present in the household for an extended period, she did a bit of travelling of her own, heading to New Orleans with a friend. The kids vividly remember the times when their mom would leave overnight for any reason, mostly because of their dad's utter incompetence in the kitchen.

"He could burn strawberries," Colin told me.

"Happily," said Nancy. "The neighbours took pity on us and started bringing tuna casseroles to the house. So we had tuna casserole for breakfast and dinner and then dad would put the leftovers in our Thermoses to take to school for lunch. Uuuuugh!"

Janet added, "Yeah, it took me decades to start eating tuna casserole again."

Doris was genuinely worried that young Colin, who was ten the year his dad was swanning about the Richmond Gulf in the dead of winter, was not bonding with his father. Knowing that many mothers have a special bond with their daughters, she encouraged Craig to volunteer to be a Scout leader and to do more with Colin one on one. Often, Colin and Craig just put a canoe on the roof of the Suburban and went fishing on some of the nearby lakes on the Frost Centre property.

On one of these trips, they drove a little way up Highway 35 and portaged to one of the secret lakes that Craig knew was full of willing speckled trout. They fished for the day, just the two of them. Colin doesn't remember if they caught any fish or not. When it came time to head home, they paddled back to the portage, and Colin, as he'd learned to do, picked up the lunch pack and the PFDs. His dad said, "You go ahead. I'm just going to lash the paddles and will be coming along behind you." Off Colin toddled, but instead of following the portage all the way back to the road where the truck was parked, he took a wrong turn and ended up lost in a swamp.

Craig finished the portage, unlashed the paddles, and loaded the canoe onto the truck. There was no sign of Colin. Craig assumed that he might have just taken it upon himself to walk home along the side of the highway.

All Colin remembers is being scared because he'd gotten turned around. But he kept his wits about him. He listened and heard the sound of traffic, so he walked toward that and eventually made it out to the road, but the truck and his dad were gone. His dad, as he later learned, had simply loaded the canoe and gone home, unable to locate his fishing partner. At home, he found Doris cooking snacks for neighbours who were coming for an evening bonfire. As she shooed Craig away from the food, she asked where Colin was.

"I lost him back at the portage," said Craig. "I don't know where he is. Is he here?"

"What? You left him where? I suggest we get back in that truck and go find him, *now*," barked an exasperated Doris.

And with that Doris sent Craig back to the fishing spot, where he found Colin sitting on the side of the road exactly where the truck had been parked, tearful but more or less unscathed. That was bonding, Craig Macdonald style.

Colin knew, as did his sisters, that doing stuff with his dad would always be an adventure. It didn't take any of them long to realize that when their dad asked them to sit by a tree at the end of a portage while he went to look for a trail, it was *always* better to go with him. It would likely be a slog to keep up, but they'd always learn something, and it was far more fun than waiting and waiting, sometimes for hours, on their own.

Looking back on growing up at the Frost Centre, Colin recalls another near-miss father-and-son adventure:

> One spring, Dad got really excited about going canoeing. The ice isn't out. There's just a moat around the shore of

> the lake. So he says, "Colin, let's get the canoe and get out there." So we get out the [Chestnut Canoe Company seventeen-foot] Cronje. We take it off the hooks. We put on our life jackets. And we start paddling along the shoreline between the ice and the land. We make it to a point a ways down the shore but then we saw this ten-foot lead in the ice that goes across to Comak [St. Margaret] Island. Dad says, "Let's go to the island."
>
> So we paddle across to the island. And when we're coming back, the wind shifts and the crack starts closing up. And my dad's yelling, "Colin, the ice is closing in on us! We're going to get crushed." So we start paddling. He was just giving it! I was small. And I could see this ice closing in on us. We almost made it, but the ice closed in on the back of the canoe. It actually crushed us. I remember it lifting me up onto the ice. And somehow we pulled ourselves with our paddles back over the last rotten ice to the moat. It would have been the end of Craig Macdonald if we were two minutes slower. This was something. He's yelling, "Paddle faster!" It lifted us up. We were out of the water on the ice. Just in the end, my paddle was hitting the candled ice, the weak ice, and he was getting crunched by it.

Bonding with Dad happened occasionally at home too. Colin had a friend over for a sleepover one weekend, and in honour of this, Craig decided to do something he rarely did, which was tell the boys a bedtime story. In the spirit of the T-Bird avalanche ghost story I had heard from Craig as a young camper at Kandalore, but having had a couple of decades of experience with Anishinaabe storytellers, Craig decided to introduce Colin and his friend to Baagak, the flying skeleton, a dark spirit — perhaps of an ancestor who froze to death from starvation. Baagak often came with winter storms, and his exploits contained lessons for children about toeing the line. Craig had heard various versions of this story through his travels and his interviews, and he had likely encountered Baagak scholarship when reading the work of anthropologists like Frank Speck and Irving Hallowell. History doesn't record which Baagak story Craig told that night, but we do know that at least one of the listeners wet his bed out of abject terror and didn't sleep a wink … and Craig was instructed in no uncertain terms that there would be no more Baagak stories told in the house.

Like the kids, Doris came to the conclusion that being with Craig was better than not being with Craig. In fact, she loved and still loves travelling with him, her memories of the laundromat at Longlac always a reminder of what can happen when Craig heads off on his own. But through the 1980s, her health continued to oscillate and her struggles with ulcerated colitis intensified to the point that her doctors recommended colostomy surgery as the only sure way to ease the pain and discomfort. So in February of 1988, Doris had the surgery and was in a Toronto hospital for so long that the kids ate a *lot* of tuna casserole. Although the post-operative recovery was chaotic — "I nearly died from the surgery," she told me. "I was so weak that I literally could not pick up the telephone" — Doris did survive, and she found new resolve in her restored health.

With life a little slower during her recovery at home and with Craig picking up the slack, Doris decided that for their family vacation that summer, they should all pile into the Suburban and drive to Labrador, a trip she'd been thinking about for years. She had heard about the Pearly Gates, massive columns of the beguiling mineral labradorite, at the mouth of the Fraser River, not far from Nain. They would go as a family, following *her* agenda for a change, and behold the summer sun energizing this geological spectacle. It was a family trip that, in the fullness of time, would make *National Lampoon's Vacation* seem like a bespoke five-star experience!

That August, with the blessings of her doctors, Doris and Craig packed eleven-year-old Nancy, nine-year-old Janet, and eight-year-old Colin into the family vehicle with all of their camping gear, clothes, coolers, and snacks, and headed to Quebec. At Sept-Îles, they loaded the Suburban onto the Quebec North Shore and Labrador Railway and settled into the lone passenger car in a gang of empty gondola cars returning to Labrador City and Schefferville to pick up iron ore. The plan was to take the truck off the train in Esker and drive over one of the most ambitious cart tracks on the planet to Churchill Falls. From there they would continue to Nain by coastal ferry.

To Craig's absolute delight, he met Innu Elder John Einish on the train, and that was the last the family saw of him for the entire ten-hour journey. Doris, glad to be out and about on the road, kept the kids fed, and she entertained the three of them by teaching them how to play euchre as they trundled north through the thinning spruce forest.

Janet picked up the story from there:

> We arrived at Esker, population two (but both were on vacation), late at night. Mom had to back the truck off a whole row of flatbeds with only narrow metal plates between them. It was nerve-wracking. (Mom is a better driver than Dad from her years on the farm driving tractors and trailers.) Someone had a trailer and it broke through one of the plates and had to be jacked up. Eventually we made it off and found the first pull-off and made camp, using the headlights of the truck.
>
> In the a.m., we discovered that the battery was dead, and everyone had left towards Churchill Falls. We tried rolling the truck down a hill to start it, but no go. We were starting to worry (since there was no train scheduled for days and days and we had only five days of food and water) but THEN, a HUGE transport truck came along and we flagged it down. The Suburban almost bounced getting a jump from that truck!
>
> We drove for *days* at 20 kph. The "road" was just the tops of eskers levelled off and snaking toward Churchill Falls. There was NOTHING and we saw NO ONE for days (now that I think about it, what did we do for gas?) except the odd pile of tons of caribou bones from a hunt, seasons before. We did meet some grey jays though, that liked eating bits of bannock off the tops of our heads. We ate bannock because bread would not have lasted. Very memorably, Dad let us sit on his lap and "drive."

> We eventually made it to Churchill Falls and rolled into the town pool. They were closing the pool but took mercy on us and let us go have showers. I recall visiting some old wooden cabin with a tall pole. We thought it was a flag pole but Dad blew our minds by telling us it was for finding the cabin when there was a lot of snow and the roof was covered right up. I will say that visiting the actual Churchill Falls and getting a tour of the hydro dam was a HUGE influence on my career.[8]

Having taken the ferry north up the Labrador coast, in Nain they found outfitter Henry Webb, who agreed to take them by boat to a beach campsite up the Fraser River fjord from which they could walk to the Pearly Gates. Along with two young female teachers from Toronto, the five Macdonalds loaded into Henry's boat for the journey along Labrador's rugged shores. While Doris and the kids might have been scanning the shore for black bears or caribou or just taking in the scenery and the saltwater smell of the sea, Craig was doing what Craig does, capturing the landscape he was seeing with a photographic eye and mapping it in his head onto stories, lore, and other learning about the place, creating knowledge that would be available for instant recall for years to come. As it turned out, that cataloguing of the landscape's features between Nain and the Pearly Gates beach campsite would be critical in surviving this part of the family vacation without catastrophe.

Henry dropped the family and the teachers off at the campsite and returned to Nain, agreeing to pick them up in a couple of days. Planning to hike to the Pearly Gates the following day, they camped on the beach. That night, Doris was overcome with pain in her abdomen. She could not stay still and lying down was not an option. Even though the blackflies and mosquitoes were horrendous, the only relief she could get was to walk in the Labrador twilight and roll on the sand when the pain overtook her.

By morning, they all realized they had a full-blown emergency on their hands. Doris needed medical help. There were polar bear tracks on the beach. They had no gun or bear spray, no map, and, most importantly, no way of communicating with Henry to ask him to pick them up.

In Nain, Labrador, outfitter Henry Webb readies the boat for a trip up the Fraser River fjord to experience the labradorite deposits at a place called the Pearly Gates. Left to right: Henry Webb, Craig, Nancy, Janet, Colin, and Doris.

At Craig's instruction the kids set three fires and wrote "S.O.S." in big letters made of seaweed to draw attention should anyone come by this isolated corner of the boreal world. But, given what was happening with Doris, Craig realized that *hoping* someone might fly or boat by and see their distress call was not a good option. A better alternative for getting help quickly was to hike back to Nain, twelve miles over some of the roughest terrain in Canada, land that he had only seen on a map and from one pass up the fjord in an outfitter's boat. And that's exactly what he did. Walking along the shoreline was out of the question, he reasoned, because the rock of the rugged mountains dropped so sharply into the sea. So with no map except the one in his head and no GPS, with just an axe, matches, a raincoat, and a bug hat, he left the kids to look after their mother and struck out for Nain. History does not record where the teachers were or what they thought about the plan. Here's how he described the hike in an email:

> Worst black flies of my life. So numerous that I could not see my route when walking into the sun because of the glint of black fly wings. Decided to climb up out of the scrub trees as soon as possible for better navigation, ease of walking and wind to reduce the number of black flies. The strategy worked, but I still had a hazy cone of black flies extending downwind twenty feet from me.
>
> Did not exactly know where Nain was. I was concerned that I might make a descent into the wrong bay. For this reason I kept to the high ridges paralleling Fraser River Bay but with a southern view to watch the passage of bays and gullies on the south side of the peninsula until I was certain of reaching the correct bay.
>
> On top, the weathered and spalled boulders on pedestals were something I had never seen before. I had to partially circumvent two small crystal clear lakes. One had beautiful purple shafts of light beaming up from a bottom of labradorite. Although very tired when I descended the final hill to Nain, I was happy that I did not get lost and was able to find Nain on the most direct route possible to get help for Doris.

Happily, by the time Craig had roused Henry and they'd sped back up the fjord in Henry's boat, Doris was feeling much better. Still, with this scare, it was Doris who suggested that she return with Henry to the clinic in Nain and that Craig and the kids should still hike with the two teachers to see the Pearly Gates. Janet didn't want to leave her mom, so she and Doris returned to Nain. Nancy and Colin stayed with Craig, and, indeed, they saw the Pearly Gates without the one person in the family who'd had her heart set on seeing this place. Doris confirmed in the clinic that the most likely diagnosis of her pain was not her ileostomy but a kidney stone, which she likely passed on the beach.

Holidays with Craig Macdonald were always an adventure and not for the faint of heart. But Doris would have it no other way. Years later, in a long letter to me about Craig, Janet wrote, "Even though Dad's style of adventure is not what Mom would have originally chosen for herself, she took to it so well and enjoyed travelling the countryside and seeing things others never get to see. It was Mom who suggested Labrador as a trip. It was Mom that wanted to go to the Yukon. It is Mom suggesting that Dad take the grandkids out winter camping in the backyard. It is MOM now who is the one planning the world travel and threatening to go without Dad when he hums and haws."[9]

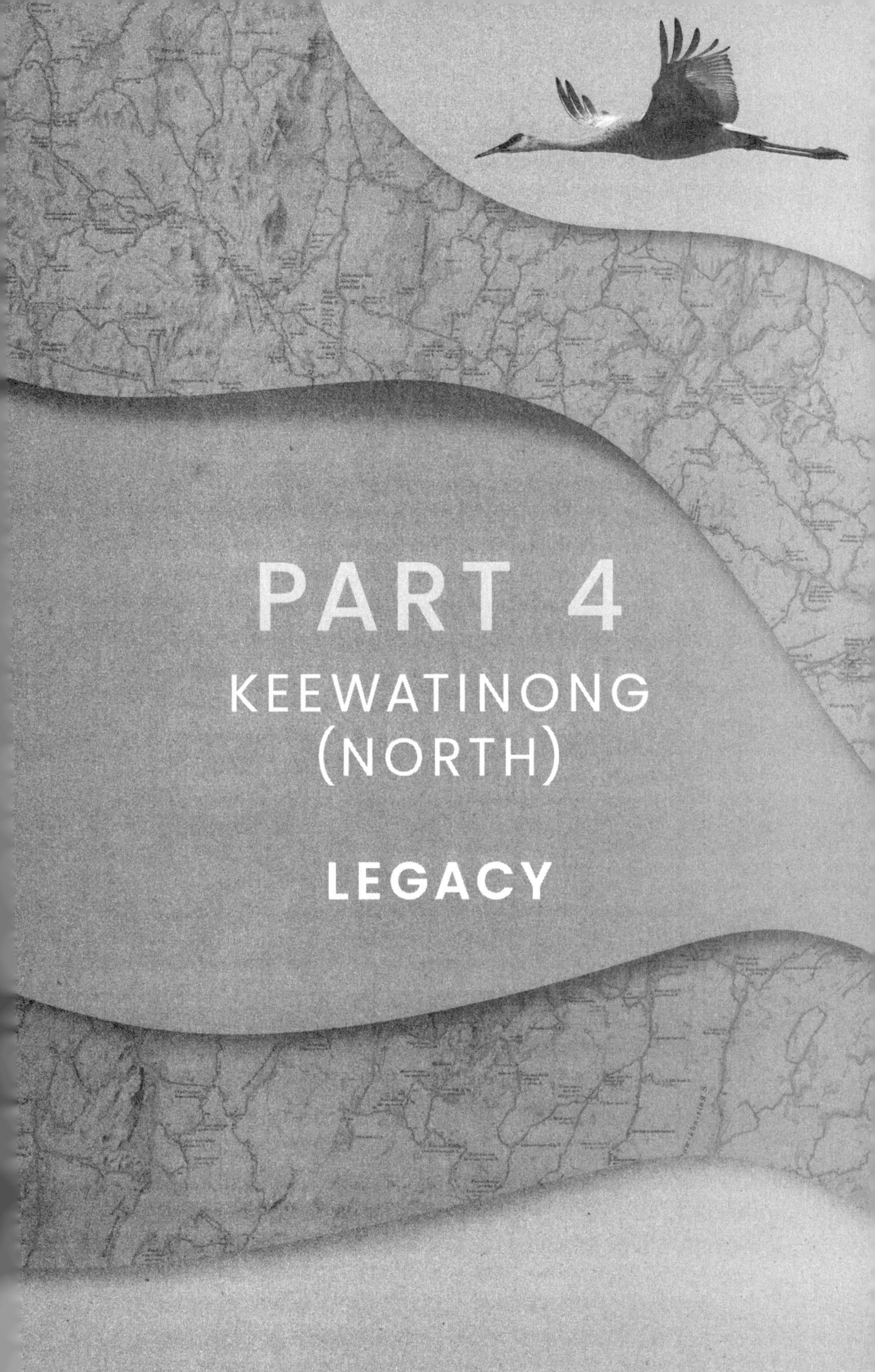

PART 4

KEEWATINONG (NORTH)

LEGACY

From Justice Steele's Decision

ATTORNEY-GENERAL FOR ONTARIO V. BEAR ISLAND FOUNDATION et al., Potts et al. v. Attorney-General for Ontario, 1984.
BETWEEN:
The Attorney General for the Province of Ontario, Plaintiff — and — The Bear Island Foundation and Gary Potts, William Twain and Maurice McKenzie Jr. on behalf of themselves and on behalf of all other members of the Teme-auguama Anishnabay and Temagami Band of Indians, Defendants.
AND BETWEEN:
Gary Potts, William Twain and Maurice McKenzie Jr., on behalf of themselves and on behalf of all other members of the Teme-agama Anishnabay and Temagami Band of Indians, Plaintiffs by Counterclaim — and — The Attorney General for the Province of Ontario, Defendant by Counterclaim....

The basic dispute is whether Ontario is the owner of certain lands, free of any aboriginal rights claim by the Indians, or whether the band or registered band has aboriginal rights in the lands that prevent Ontario from dealing with the lands until those rights are properly extinguished.

The defendants referred to the land claim area as "Ndaki Menan." I refer to it as the "Land Claim Area." ...

The evidence called at trial was extremely lengthy, far-ranging and comprehensive. The trial lasted 120 days and well over 3,000 exhibits were filed....

Like other evidence, oral evidence is not always accurate.... Oral history may also be contradicted by available factual records. The same is true of the Temagami area. This shows that while oral evidence must be weighed like other evidence, consideration must be given to the faultiness of human memory.

I feel obliged to comment on how disappointed I was that there was so little evidence given by the Indians themselves....

Throughout the trial I had an uncomfortable feeling that the defendants, in presenting their case, did not want the evidence of the Indians themselves to be given....

If a white person, or non-Indian, gives evidence as to oral tradition, this testimony is admissible generally only where the declarants are dead. In the present case, Mr. Conway, Mr. Morrison and Mr. Macdonald gave evidence as to what they had been told by Indians. In many cases, the supplier of such information is still alive and in other cases is not identified. This must be borne in mind in determining the credibility and weight of the evidence. I make this comment notwithstanding that to a degree all three of these gentlemen are experts in their own particular fields....

In my opinion the plaintiff is entitled to the relief claimed and the defendants' counterclaim fails....

Craig Macdonald was a witness who has done extensive work on canoe routes and recreational matters, as well as having general knowledge of biology and fish and game maintenance. While he is very familiar with the Temagami area in general, it was obvious from his comments that he viewed the court process as being rigged against the Indians and that, from his two references to "we" in the context of being part of the defendants' team, he was biased in favour of the defendants. He, Thor Conway and James Morrison were typical of persons who have worked closely with Indians for so many years that they have lost their objectivity when giving opinion evidence.

I accept from Mr. Macdonald's evidence that maple sugar was a resource used by the Indians in more than one location in the Land Claim Area, and that there were herbs and berries in various places. This indicates that the Indians probably used them as a resource. His evidence also shows that the Indians used sleighs and snowshoes. This shows that they used the forest products and animals to manufacture them. I find that there was no distinctive style of sleigh or snowshoes in the Land Claim Area. His evidence also indicated that birchbark canoes were built.

The great bulk of his evidence was hearsay and much of it was obtained from Donald MacKenzie who died after the commencement of this action but before the trial. Mr. MacKenzie was one of the old people of the defendants and it is clear that an electronic tape recording was made of interviews with him. These tapes were not produced by the defendants because they claimed privilege in them. It would appear that Macdonald's evidence is not the best evidence available and therefore I discount his hearsay evidence to a great extent. For instance, when he referred to one site as a quartzite mountain, he made no reference to any mining of it. In another location he referred to vermilion but not to any removal thereof. However, in his hearsay evidence he indicated that pipe-stone had been quarried at one location in the south end and vermilion at another. It may well be that the aboriginal Indians did quarry pipestone or take vermilion but this would be in limited quantities and would be for their own use, and therefore usufructuary....

Craig Macdonald gave extensive evidence with respect to the Indian names for a large number of minor and major geographical locations within the Land Claim Area. He stated that he started amassing this information prior to this litigation, and that he believed that the information was accurate. The information came from local Indians, but none of the latter were called to give evidence to support him. I believe that Mr. Macdonald has done an extremely capable comprehensive documentation of these names, but I am of the opinion that the evidence is of little value to this trial. Mr. Macdonald himself indicated that the oldest of the names, which he sometimes expressed as "a long time ago," dated from the eighteen hundreds and, in light of his answers, probably from no earlier than 1850. From this, it is clear that most of the names are of more recent origin, and clearly could be names passed from grandfather to father to son, and do not indicate long-time occupancy of the land....

Macdonald also gave extensive evidence as to the canoe routes within the Land Claim Area, and stated that in his opinion the Land Claim Area formed a natural geographical area that was consistent with the position that there was one group of people living within the area. I do not agree.

I felt Mr. Macdonald was extremely biased in favour of the Indians. Most of his evidence-in-chief was given from notes, which is understandable because of its detailed nature. However, in cross-examination, he appeared to have difficulty following the documents put to him to be considered. When asked about the Nipissings that he interviewed in the nineteen seventies with respect to territories and canoe route access to the Wanapitei area, without any apparent reason he exploded and stated that the court rules were loaded against the Indians. At another time, he mentioned that he was so worked up that he could not remember a person's name. Still later, when questioned about names of old geographical points, he stated that many informants of Bear Island can give the name of why Rabbit Lake was so named, and then stated, "This will become more evident when *we* start pouring in the documents." All of this indicates that he was clearly emotionally involved and a partisan witness.

He referred to numerous entry points into the Land Claim Area and referred to two or three alternative routes going north out of the Land Claim Area. However, concerning the westerly side where the defendants do not want to show any connection with the Wanapitei area, Mr. Macdonald, having indicated that there were perhaps five or six accesses, in my opinion wrongly concluded that all the accesses were bad. He also said that the routes in that direction were not clearly inter-connected like veins or arteries, whereas inside the Land Claim Area they were, and in support he presented a water-fill map, Exhibit 19-7, which he said showed the absence of lakes between Wanapitei and Temagami. In my opinion, it also basically shows the absence of lakes between the northern part of the Land Claim Area and the southern part.

He also gave evidence with respect to Kelly's portage, which was a two and a half mile portage over some high ground that he said would be the best access point between Lake Wanapitei, and the Sturgeon River and general Temagami area, but he said that this was very difficult and would not have been used as a normal route. From other evidence, it is clear that this is the very area or portage that was used by one of the claimed ancestors of the Temagami band, Old Barbu, who was found hunting there in 1839. It is also clear that the Indians in those days would not have considered a portage of two miles or two and a half miles a *major* portage. From Exhibits 19-1 and 19-2, prepared by Mr. Macdonald, one can see that there were many portages within the Land Claim Area that are of that length or longer. In particular, evidence was given from historical documents of a portage between Lake Mattawagami and Sinclair's Lake to the north of the Land Claim Area that was some seven miles or more in length and over which old women regularly passed. While a flat portage is obviously easier than a high portage, there is no question in my mind but that Mr. Macdonald was being highly defensive of the Indians' position, over-emphasizing certain factors to make certain that there was no connection between Lake Temagami and Lake Wanapitei. He exhibited extreme bias in refusing to acknowledge the obvious with respect to the Skene map based on information obtained from Chief Tonene of the Temagami band in 1883, and in particular stating that the lake shown on the Skene map was likely Lake Wawiagama, thus rendering his position ludicrous and of no use. He admitted that his evidence with respect to snowshoes and toboggans failed to accomplish his goal of indicating some special band difference from other bands. I find that his evidence does not prove the existence of a unified band nor determine the area covered by any such band....

CONCLUSION

In conclusion, the plaintiff is entitled to a declaration for the relief claimed in paragraphs 6(a), to (e) inclusive of the statement of claim. This relief is set out in the Introduction to these reasons. The counterclaim of the defendants is dismissed.

I may be spoken to with respect to costs.

[signed] Justice Donald Robert Steele in the Supreme Court of Ontario (Toronto Non-Jury)

Released December 11, 1984.

Lifeline IV — Moving On

AS IF JUSTICE STEELE'S STINGING REBUKE AND REPUDIATION OF HIS work in his land claim decision and the OMNR's stalling of the production of his map were not enough to kill Craig's joie de vivre, while he and the family were in Labrador, all was not well at the Frost Centre. It was a big and aging facility and, although centrally located in the province, it was almost too far from anywhere. The original vision for the site, with its 24,000 hectares of trails and wildlands and accommodations for two hundred people, was a multipurpose outdoor public education centre for youth and adult outdoor enthusiasts. But the numbers of workshops, gatherings, conferences, and busloads of schoolchildren were declining. In addition, the Frost Centre was supposed to be the principal training facility for the OMNR and the Ontario Public Service, including the Ontario Provincial Police and the Ministry of the Environment. But the facility was not being used enough to sustain it. That problem, coupled with the increasing cost of maintaining the crumbling physical infrastructure of the big campus on beautiful St. Nora Lake, was catching the attention of the government bean-counters at Queen's Park. By 1988, the writing was on the wall for its eventual closure and/or sale, which meant that mid-career employees like Craig, who was just forty-two at the time, were encouraged to consider other employment within the OMNR.

During the ten years he had been at the Frost Centre, Craig had settled into his recreational specialist's role, which allowed him to spend the days outside teaching, maintaining trails, or doing any of the other thousand things that a handy outdoorsman might do, while at night and on weekends he continued with his research. But even he could see that change was in the wind. The Frost Centre was likely not going to survive much longer due to cutbacks. And, as an interim step, OMNR wanted to redeploy Frost Centre staff to specialist positions in other areas of the organization. With the counsel of Doris and his Frost Centre boss, Jim Barker, Craig reviewed the list of alternative positions and settled on an opening for district supervisor of the Bracebridge District of Ontario Parks, working out of the Bracebridge Resource Management Centre. The position would last until this district was downsized and incorporated into the Parry Sound District.

In his new role, Craig was managing operations for several parks in the Muskoka area including Mikisew, Arrowhead, and Hardy Lake, as well as overseeing activities in the OMNR's Crown land program. This job was not a great fit for Craig. Whereas at the Frost Centre he actually lived at his place of work, the new job required a drive, either north to Dorset and west on Highway 117 or south to Carnarvon and west on Highway 118 over to Bracebridge. Either way it was a commute for the better part of an hour each way. Although he did pick up a shovel and a Pulaski to make some modifications of the entrance road at Arrowhead Provincial Park, to solve a problem with winter access, which was really not in his job description, and he did fire up the dual-tracked single-ski Ski-Doo Alpine to groom the Bracebridge Resource Management Centre ski trails, as he had done at the Frost Centre, these were just ways to avoid the tedium of paper-pushing at the office. Craig hated this aspect of the job, a situation that was not lost on his bosses at the OMNR.

There was much to recommend Craig Macdonald for employment in the natural resources and parks sector, not least his penchant for hard physical labour and his ability to run a track setter. Colleagues and strong young summer interns loved working with him in the field. They called him Craig-o-matic[1] because he was a machine when it came to hard work, from dawn to dusk. His capacity for hard physical work was legendary. He could site a trail with map and compass. He could build a bridge. He was a licensed blaster who, with a star chisel or diamond bit or just a pack of well-placed mud and some dynamite, could crack and move boulders that were in the way of a beautiful ski runout at the bottom of a trail. His bush skills were impressive — shelters, fires, snares, traps, axemanship, navigation, snowshoeing, canoeing, ATVing, snowmobiling — in all seasons in all conditions. He was trustworthy. He was loyal. He was a great teacher and raconteur. And he *loved* being in the bush. Oh, and he spoke Ojibway and knew the Indigenous names for just about every tree, plant, fish, insect, trail, and weather or water condition he might encounter.

In recognition of his unique value to the organization (and possibly as a favour to itself, so that it could install a more management-oriented soul in the Bracebridge job), the OMNR created a unique position for Craig, recreational specialist in Algonquin Provincial Park, a job that gave him free-rein responsibility for trails, campsites, and anything else his fertile mind might dream up to enhance the recreational experience of the entire park.

In this new role, as the only employee with this job title and description, Craig didn't really fit into the established administrative structure of the park, but that didn't seem to matter. He was classified as an "interior" — at least that is where his budget was housed — so he did not work with campground employees. He answered directly to the superintendent of the park, but he was also on the park management team that dealt with big issues in the park, which worked out of an office at the East Gate where he was almost never to be found. It was a win-win situation for Craig and for the OMNR.

At the time, however, there was a perfect task for Craig in the offing. Park personnel had determined that the daily quota system set out in the Algonquin Park Master Plan, approved by the provincial cabinet in 1975, was defective. A new approach was required. Up to this time the park had no field information on which to establish the interior carrying capacity for Algonquin Park. Craig's task was to make this determination with fieldwork in Algonquin. The task was to check the shorelines for camping potential and to accurately locate existing campsites on every lake and every island on every canoe route in the park. This was a massive undertaking that was also used to update the Algonquin Park map with campsite locations.

Work on the water continued regardless of weather right to freeze-up. Craig and his assistant, George Oram, would often return to their interior cabin with their clothing caked with ice. After freeze-up, work continued with a cold, unheated M.A.S.H.-style glass-bubble Bell 47 helicopter on floats until the snow got too deep for an accurate shoreline assessment.

There was so much to be done. Operations Manager Ron Speck sought Craig's help to figure out how to turn the old Booth railway line into a bike trail. Building dog trails, snowshoe trails, and hiking trails; maintaining portages; finding lost campers; negotiating with local First Nations and forestry companies; building new campsites; decommissioning old sites; mapping Indigenous trails and sacred sites; organizing commemorative treks across the park — all of these were within the purview of Algonquin's new recreational specialist. Craig had the freedom to roam and a park-wide brief to follow his bliss in the fresh air. In later years, because he was often working alone in remote areas of the park, they decided to put a GPS tracker on him so they could find him in the event of a mishap.

The planned closure of the Frost Centre also meant that in 1990, for the first time in their married life, Craig and Doris (and the children, now fourteen, eleven, and ten) had to shop for a house to purchase. By then, Doris had gone back to teaching at Dorset Elementary School, just eight miles north. Happily, they found the perfect place just nineteen miles up the road — a sprawling four-bedroom bungalow with a full unfinished basement — in the village of Dwight at the junction of Highways 35 and 60. It was almost exactly halfway between Huntsville, where the kids went to school, and Craig's work in Algonquin Park. And, as a bonus, the property on Sale Road backed onto wild lands along the Oxtongue River where they could ski, snowshoe, walk, bike, and build forts right from the back door.

Craig was never much for creature comforts, and the best part of the new house for him was the unfinished basement, more than twice the size of his basement grotto at the Frost Centre. Here he could keep his research papers, books, sleds, tents, tools, snowshoes, maps, cold weather gear, paddles, and other miscellaneous Craigian flotsam. With the map's production still stalled, he was now hell-bent on writing a book about all the winter travel technologies and techniques he had been absorbing from his research. The only tool required for that project that wasn't in his basement space was a computer. Nancy remembers her dad in those early days at Sale Road, sneaking into her bedroom in the wee hours of the morning to peck away on her word processor. As far as Craig was concerned, anything that wasn't wood-powered could, and should, be maintained by someone else in the family.[2]

But just like the basement at the Frost Centre, the basement at Sale Road became a place of singular focus for Craig, aided and abetted by Doris, who still, with a full-time teaching job, handled most of the other tasks and duties of child-rearing — running kids back and forth to lessons and practices — and keeping the Macdonald home fires burning. "He didn't really care how things happened in the household," Doris told me. "But he never interfered with how I raised the kids. Now, maybe he thought I was doing an okay job. I made the decisions. And you know what? I was raised that way. My mom made the decisions. The only time my dad ever interfered was — if he ever got involved — well, I knew I was really, really in trouble. Because I was raised that way, I felt that's the way we should run our relationship. I felt Craig's out making the money and my job is to run the house."[3]

"Normal" is what happens by rote or habit in any family, and the Macdonald household was no different than any other. About the parenting mix, son Colin said this:

> You have to remember that he was so busy with his research that my mom was basically a single mom with a dude that was around a bit. So him taking off on a trip wasn't really that big a change in lifestyle for us. He would disappear and you wouldn't really know he was gone. My mom was there to support him, but he gave a lot of leeway to my mom, if that makes sense. And that leeway sort of spilled into everything the guy did. He had a lot of leeway with us kids. I'd be going on a canoe trip and basically his only advice would be "Don't wreck the canoe." Everything else is game, but don't wreck the canoe.[4]

Through their genes or by osmosis, guided by the sage teachings of their mother with occasional colour commentary from their dad, all three of the Macdonald siblings thrived at the Frost Centre and through the move to

Dwight. They were all active and engaged in ways that reflected their interests and personalities, almost in spite of their dad's semi-maniacal passions. And, each in their own way, they excelled at what they did. Janet was good at everything she undertook, including playing the trumpet. Here's her story about her dad's reaction to her becoming a solo performer in the Huntsville High School Stage Band:

> When I was in the tenth grade, [in one band recital night at school] an entire song was my solo with the band backing me up. It was a big deal. It was a big honour. I was super excited. Dad drove me to band. I had to be there half an hour early for the concert. And he disappeared. He never showed up. Because, you know what he did? He went to Huntsville Library and got lost in some book. And then he picked me up late, after the concert was over. I was hurt by this at the time but it's just classic Dad. He just didn't think it was important.[5]

Nancy, the eldest, said this:

> There's no question that Dad loves us and we love our dad. He's proud of us. There's no question there. He never held us back from anything, but if our interests didn't align with his interests then he had no interest. For example, I played basketball in high school. My dad was not particularly interested in basketball. We never watched basketball games together. My mother would ask him to take me to basketball games at school, because I had to be driven. I'd say, "Dad! Did you see that hoop?" And he'd say, "No, Nancy, I was reading the paper." And it didn't bother me. It is what it is. It's an ongoing joke. So he supported us in the sense that he drove me to the game.
>
> But if I had turned around and said, "You know, Dad, I want to go buy a canoe," or "Dad, I'd like to go on a

> snowshoe route somewhere," he would spend hours and hours and hours talking to you about it.
>
> It didn't interest him to discipline his children and ensure that they were getting their homework done. If we weren't doing our homework and we were failing, usually what would happen is my mother would come down on us like a hammer or my mother would turn to my father and say, "Craig, you are going to have to say something to your children." At that point, Dad would come into the room, say three words, and that would be the end of it. But you knew you were in real trouble when Dad got involved.[6]

By all accounts, however, there were plenty of opportunities for Doris and the children to accompany Craig on ventures by canoe and snowshoe to find a portage or document a trail, to locate an informant in a tepee on some distant First Nation reserve, particularly during the summer. And, for better or worse — the laundromat at Longlac was definitely notched in the latter column for all concerned, except Craig — these had as many comedic moments as they did trials.

While some parents would play Raffi's "Baby Beluga" to children in car seats, over time replacing this with car sound system selections to honour the children's choices in music, by contrast Craig had Ojibway language tutorial tapes in high rotation on long road journeys. So much so that all three of the Macdonald children now have a rich vocabulary of random Anishinaabemowin words and phrases they can toss into conversation.

With the least provocation, Craig could riff on language ad nauseam, as he could on so many other things. Nancy described an Easter dinner with her husband and three children at which her dad started explaining to his grandchildren what the word for "Easter" is in Ojibway. Which was fine, except that he continued by naming every item on the festive dinner table in Ojibway. "After a while, the kids just stared at him, their eyes glazed over," Nancy said.[7]

Nancy's response to this charming dinner table vignette may have been coloured by the memory of the time she brought a new boyfriend

Doris captures Craig and the children exploring a hand-hewn log trapper's shack in Labrador. Left to right: Janet, Craig, Nancy, and Colin.

home for dinner at Sale Road, a boyfriend who happened to be from Saugeen First Nation. On being introduced, Craig heard the beau's surname and then launched into an extended soliloquy on his family history. And for the rest of the meal, Craig insisted on speaking to the visitor in what he thought was his first language. "This might have been fine," said Nancy, "except that my boyfriend didn't speak a lick of Ojibway. Dad was so excited that I'd brought someone home with the same passion as him. I think my boyfriend was interested in his culture and so was interested in having this conversation with my dad but, you know, it's not the coolest dating story."

In 1993 a delivery truck rolled up to the Sale Road house to deliver Craig's portion of the first and only print run of the *Historical Map of Temagami*. He had imagined this moment many times since the very early days of his research in the late 1960s, when he first had the idea of making his own map. And, after that exhilarating moment in 1985 when he spoke to the Ontario Geographic Names Board and received their ringing endorsement of his work, followed by their commissioning the completion of the map through the auspices of the OMNR, he might even have pictured the day when he would return to each of his informants to present them with a signed copy.

But nearly a decade had ticked by since then. His bitterness with the production hassles and delays lingered. He was still working for the OMNR, but he was in a different job. New management at Queen's Park seemed to value the map even less than the group who had tried at the last minute to squelch the project altogether. The analogue map-makers in the Cartographic Division of the Office of the Surveyor General were retiring and their technical prowess and artistry was giving way to the recently graduated GPS whiz kids who spoke only digital cartographic languages.

And yet, what came out of the brown-paper-wrapped packages he took off the truck and stowed in the basement — lit only by bare incandescent bulbs — was a totally arresting three-by-four-foot portrait of the dreams and deep land wisdom given to him by scores of Anishinaabe Elders — mapped knowledge and *nastawgan* that had been dismissed by Justice Steele at the Ontario Superior Court of Justice, by the Court of Appeal for Ontario, and, most recently, by the Supreme Court of Canada. Bittersweet doesn't begin to touch the quiet intensity of this moment.

Craig's first impulse on receiving the shipment was to put in motion a plan to visit as many of his informants as possible so that he could sign and present them with a copy of this priceless cultural artifact he had pieced together from generations of lived experience on n'Daki Menan, which until now had lived only in the heads of Elders who were slowly taking it to their graves. But the trial had left the Teme-Augama Anishnabai fractured and disillusioned. Many were not sure what to think about their language and traditions or their storied past, or whether to even care about the map and how it might relate to the illusions and confusions of today or the fortunes of tomorrow for their children and their children's children.

Whether they had or hadn't signed on or "adhered to" the Robinson-Huron Treaty of 1850, the court said they had and thus had given away, or "extinguished," their ancestral land rights. And since that time the Deep Water People of Temeeaygaming had literally lived a "war of the worlds," theirs versus that of successions of settlers who would wheedle, encroach, and steadily pick away at life as it once had been for the Teme-Augama Anishnabai. The HBC. The school. The train. The railway trappers. The Forest Reserve. The loggers. The diseases. The restrictions. The regulations. The miners. The cottagers. Through each of those waves of outside influence, the First Peoples of Temagami had found ways to absorb and accommodate the changes, but invariably, inevitably, they gave up more than they ever gained from "progress."

In 1928, ten years before Andrew Couchie and Larry Turner inadvertently tuned into Orson Welles's fictional story of their lives on their fancy portable radio at Solace Lake, the government of Ontario was attempting to charge the people of Bear Island rent, claiming that they really had no right to be there.[8] In fact, for decades by then, the Teme-Augama Anishnabai had been asking for land to call their own. As far back as 1870 the band had petitioned the government for reserve lands. And since then, the governments of the day had agreed, up to a point, even going as far as surveying a reserve at Austin Bay on the southeastern corner of Temeeaygaming in 1884, before taking it back again some years later because the timber in that area was simply too valuable. Indeed, even when all 1.8 square miles of Bear Island were finally set aside as federal reserve lands for the Teme-Augama Anishnabai in 1971, the issue of other reserve lands in the area was not resolved. Today, the issue of other lands in the area for exclusive use of the Deep Water People is still not resolved.

The one thing the trial did establish in favour of the defendants, however, was that although they had accepted treaty payments and thereby extinguished their rights to n'Daki Menan, the government still had an obligation, a "fiduciary responsibility," to resolve the issue of mainland territory for the people of Bear Island. Unfortunately, in 1975 Chief Gary Potts, attempting to define his people as something other than "*Indian Act* Indians," created with the help of lawyer Bruce Clark a legal organization that included the so-called Status Indians of the Temagami First Nation living on Bear Island, but also counted non-status, Métis, and other people with a connection to the Bear Island community. Following Judge Steele's decision, it was that larger group — formally known as the Teme-Augama

Anishnabai, as distinct from the Temagami First Nation living on Bear Island — who entered into negotiations with the provincial and federal governments to settle the land issue once and for all.

Following the Supreme Court decision, there was energy to get this resolved, fuelled by anger and disappointment. The Anishinaabe had been talking about this for generations now, back and forth with government after government, and by the early 1990s, about 130 square miles of land on the east side of Temeeaygaming had been "set aside" in anticipation of the eventual signing of a final agreement. There would be room for building a mainland community at Shiningwood Bay, and although there would be nothing in this tract akin to the natural resources of the whole of n'Daki Menan, which had sustained families for generations, these lands would provide a base for traditional activities and would be there for the Teme-Augama Anishnabai to manage as they see fit.

Sadly, by the time this draft settlement agreement came to be ratified by the people in 1993, there was sufficient mistrust in the leadership and misinformation about what was actually going on that the community as a whole was divided. The Teme-Augama Anishnabai, the broader group living off the reserve, voted overwhelmingly to accept the deal, but the Temagami First Nation, the people of Bear Island, turned it down.

It was into that unfortunate swirl of intergenerational trauma and internecine enmity, anger, confusion, and disappointment that Craig Macdonald's map finally landed. In the shadow of the vote, Craig went to Bear Island and did a presentation for the community. It was all very polite. Afterwards, at a table in the rec hall, he personally inscribed maps, one by one, for each of his informants who were still alive, laughing and visiting with them as old friends would. For the families of those who had passed, he signed maps in their departed Elder's memory. But it would be years before the true value and significance of what the Elders had given Craig would be publicly appreciated and recognized.

For a more recognition-oriented person, the muted response of the community to the publication of the map might have been disappointing, but Craig Macdonald is just not wired that way. His mission in returning to Bear Island was to honour, appreciate, and say thank you to everyone who had contributed to the creation of the map. As he personalized copies of the map — many of which were laminated and hang to this day on walls throughout Temagami and beyond — he was doing what he felt he needed to do, which was to share the final physical embodiment of nearly a quarter century of conversations about routes and place names, Traditional Indigenous Knowledge that had been given to him freely by so many Teme-Augama Anishnabai. And even if the OMNR had no intention of doing anything with his map, bringing it back to Bear Island completed the circle of trust in which the project had been conceived in the first place.

• • •

And, truth be told, all the while that the map project had been suspended by mysterious forces within the OMNR, Craig had not missed a beat in continuing the research toward his next and equally significant project, a book that would document traditional winter travel technologies and techniques, doing for them what Edwin Tappan Adney's *The Bark Canoes and Skin Boats of North America* had done for traditional watercraft technologies and building techniques. Craig knew that Adney's work was the bible of bark canoes. With the zeal of a self-anointed prophet, he was committed to writing and illustrating a similarly influential treatise all about sleds, toboggans, and snowshoes.

The first version of this manuscript — and it really was a *manu*-script because it, like all his notes, was pencilled on foolscap in his chaotic longhand script — was actually a guide for people who had heard about Craig's passion for "hot" winter camping and wanted to try it out for themselves. Not surprisingly, this first abbreviated draft of what became the book was produced in 1985, in the space created in his life when the Ontario Geographic Names Board agreed to publish the map. With the map out of the basement (but never out of his mind — Craig would be the first to admit that he drove the OMNR cartographers crazy with constant updates and addendums as he learned more from his continuing interviews about the *nastawgan*), Craig was able to turn his attention back to his notes to lay out a structure for the book.

As the map languished in the catacombs of the Office of the Surveyor General, Craig poured through his interview notes and reflected on his efforts to recreate in his workshop and test on his expeditions snowshoes, toboggans, sleds, harnesses, and other winter travel accoutrements. As with the place names and trail information he had been given by the Elders, he knew that knowledge of the travel technologies and techniques was fading, as fewer and fewer Indigenous people were living and travelling on the land. Unlike the map, however, which was limited to n'Daki Menan, the traditional lands of the Teme-Augama Anishnabai, the book he had in mind spanned the full sweep of his research area, which now included first-hand experience on the land and interviews with knowledge holders from Labrador through the boreal forest to Yukon and Alaska.

In a way, the trip to Temagami in 1993 to bring the map back to the community, although necessary, took time away from Craig's basement lair where he had been meticulously sketching diagrams and writing detailed descriptions to illustrate every conceivable aspect of the winter travel technologies and techniques he had been learning about for over twenty-five years. In 1990, for example, a young Temagami outfitter called Andy Buckman, who ran a company called Northwest Expeditions and was a huge fan of Craig's, called to say that he was organizing a winter trip to Mountain Lake, just south of his cabin on Anima Nipissing Lake. The plan, he explained to Craig, was to invite Michael Paul — Kush Kush — and Bill Twain, two Elders Craig had already interviewed, into the bush so they could pass on some of the hands-on techniques that would have been difficult to convey in a kitchen-table conversation. Craig, of course, jumped at the chance to join in.

Craig had spoken to both Kush Kush and Bill Twain multiple times going back to the 1970s. But these had been in-town interviews. In fact, the last time he had talked to Kush Kush was in a hospital in North Bay, where the Indigenous Elder had just had eye surgery and was in no shape to travel. The chance to go with these two key knowledge holders into the bush, in the winter, was unparalleled! And to have the logistical support of Andy Buckman and some of the strong young guys who worked for him would potentially make this junket one of the most informative and revealing research forays ever.

Their destination was a particularly beautiful place on Pishabo family traditional territory that, by coincidence, happened to have the same name as Larry Turner and Andrew Couchie's destination back in 1938: Sawgidjeewayawgamawk, which translates to "that lake you can see when you get to the high point on the portage leading into it." To make this trip possible, Craig and one of Andy's staff hand-hauled sleds on snowshoes down Anima Nipissing Lake and over the portage into Kawwawshigamōng Sawgihaygunning (Breeches Lake). Here they set up two winter wall tents that would serve as a warm place for the Elders to camp when they arrived with Andy by dog team.

As it turned out, these lands had been clear-cut, and to add complication to the remains of the logging activity, the steep trail up to Sawgidjeewayawgamawk was littered with windfallen trees. Craig and Andy's helper got the camp set up and did their best to clear the portage, but when it came time the following day to take Kush Kush and Bill up the hill by dog team, the snow was so deep and the conditions so unfriendly that they eventually had to abandon that plan and return to the warm winter camp.

Bill Twain, seventy-three years old in 1990, was pretty spry in the step and only too happy to fudge his way through the deep snow on snowshoes. Although Kush Kush was seven years younger than Bill, he had been struggling with diabetes and vision problems and wasn't in any shape to snowshoe. Speaking to me about this trip years later, Andy said,

> My abiding memory of that trip, Jim, is with Kush in the sled and Bill in snow literally up to his waist and just heaving the sled and the team and Kush and the whole thing. Bill was a pretty powerful guy, right, and he's just heaving the sled up the hill, like six inches at a time, doing this brutal work. But he's laughing and smiling and his eyes are sparkling, and he's just laughing and saying, "This is how we used to do it, eh, Kush?" I have never ever seen a person so much in his element on the land before or since in my life. He was just having a ball! Because he was in intimate contact with the land. It was how he had grown

Bill Twain.

Michael Kush Kush Paul with his son Alex and grandson Alex Jr.

> up and what he had known, and he just loved it! It was really, really fun.[9]

Back at the tent camp the following day, Bill decided that he would show everybody how they made deadfall traps back in the days before metal traps were available. Later, Andy had a clear memory of Craig following Bill's every move and taking notes as Bill went about finding a location for the trap, gathering the materials he needed, and making the trap with nothing more than a knife, an axe, and a lifetime of know-how.

> Bill was a really energetic guy. But when he was doing a physical task he would be methodical and slow, but incredibly efficient at getting things done, cutting the pieces out. I have a sort of vision of looking at the finished trap. He had taken some boughs and put them around it. He was demonstrating to Craig exactly how the deadfall piece should be positioned so that it came down directly and evenly in contact with the lower log, so it would make a clean kill. Bill just sort of fiddling and fussing and adjusting and tweaking and trying to get it just perfect. And meanwhile, Craig just asking question after question after question after question after question, and Bill answering question after question after question. "Why do you do it this way?" "Why do you do it that way?" "How come you peeled the bark here and not there?" I remember the two of them, again, being intensely focused on the trap. Everything else, what they were wearing, or — I remember them both bent over the trap in the snow at the side of the lake. But everything else is incidental. Everything else is what two highly experienced people would be wearing in the bush in January.

Andy Buckman eventually sold the cabin on Anima Nipissing Lake, but he continued his quest to learn about Indigenous trails and travel techniques as inspired by his friend and mentor, Craig Macdonald.

> It's hard for me to express in words how much respect I have for Craig and his work. He is so meticulous, and he's so interested in learning. It's invaluable. It inspired me, in later years, to document the traditional trails and routes in the area around Darrow Camp in northern Maine. His work represents a depth of focus that's just a model for bridging between different lifeways, bridging between different cultures. I hold Craig in awe. And he does all this in a way that is so unassuming. There is no hint of any kind of a desire for his own power or his own influence. He just cares about the land, and he cares about the people and cares about preserving what we were fortunate enough to learn.
>
> I think he has been a model for many of us in thinking about all the stuff that's come up subsequently after all this of reconciliation and trying to come to grips with both the problems of the past.... But I think that part of that is also coming to grips with some of the really good stuff of the past and the extent to which people have been able to share and learn from each other and find mutual respect. Craig was a pioneer of that kind of stuff. I think he was way ahead of the curve in all this.

• • •

As Craig's reputation continued to grow after the quiet release of his map, it was as if his frustrations with the OMNR's stalling of the map project had doubled and redoubled his energy and commitment to documenting the minutiae of traditional winter travel technologies and techniques. Now, in the basement of his home in Dwight, surrounded by all manner of sleds, snowshoes, toboggans, and harnesses, and the smells of raw wood, linseed oil, babiche, pine tar, woodsmoke, and bacon (emanating from his Egyptian-cotton wall tents, which hung from the bare floor joists overhead), he would lose himself in what would be the major written opus of his life, a book he called "Traditional Sledding in North America."

In 1998, now that their youngest, Colin, had finished high school and joined his sisters in post-secondary studies, Craig and Doris decided that what they really needed for their summer vacation was to fly to Whitehorse to mark the centennial of the Yukon gold rush. Given all the yawning days, weeks, and months when Craig would work till late in the park and then come home and have a quick dinner before disappearing into the basement until everyone else had gone to bed, Doris loved — and loves to this day — travels with the man she fell madly in love with through their epistolic courtship nearly thirty years earlier. But she knew better than anyone that no matter where they went — it didn't matter — Craig's curiosity would sideline them somehow. At least this time, as they packed all their camping gear into a rental car in Whitehorse, there were no diapers, kids, or car seats to complicate the journey.

And, as they crossed the Yukon–Northwest Territories border and pulled into the Gwich'in town of Fort McPherson, Craig did not surprise his travelling companion. He needed to find Robert Francis, a Gwich'in Elder who was a renowned snowshoe maker and accomplished knower of winter ways in the lands of the Mackenzie River Delta. On the edge of town they happened upon a chap weaving along the side of the road, so they stopped to ask if he might know if Robert Francis was in town and where he lived. Indeed, although he was clearly intoxicated, the guy did know Robert Francis and knew where he lived. In fact, he offered to show them, if they would give him a lift into town. The only trouble was that with all the camping gear, there was no room for another person in the little rental car. The solution? Whether Craig asked or Doris offered — perhaps it was a bit of both — Doris got out onto the side of the road and Craig and his new tour guide headed off into town.

For a while, Craig wondered whether this was actually a good idea, because his guide took him on a tour around the town, laughing and waving to all of his friends in the manner of a royal touring a borough of serfs. But he did actually know where Robert Francis stayed when he was in town, and eventually they got there and Craig did connect with him for an interview

— but not before a quick run back to the edge of town to pick up Doris and thank her for her contribution to the connection.

The interview with Robert Francis was to snowshoe typology and nomenclature what all his conversations about place names and routes were to producing his map. Those conversations with informants in the far-flung corners of the boreal, like Francis in Fort McPherson, John Einish in Labrador, or, dare one say, Joe Bananish of the famous laundromat incident in Longlac, always added details about routes and place names to the comprehensive map in Craig's head. But it was the techniques and terminology for building, using, and maintaining snowshoes; keeping dogs and dog harnesses; and making sleds and toboggans that he would write down in his notes from these encounters. Every conversation added to the "Traditional Sledding" manuscript, which, by around the year 2000, had grown to thirteen chapters, with a fat glossary of terms and a bibliography, totalling nearly fifty thousand words and illustrated with hundreds of classic Craig Macdonald cartoon-like drawings of how everything worked.

One of his closest confidants through all of his research was an engaging bush scholar, outfitter, and canoe maker named Hugh Stewart. Craig had met Hugh in Toronto during the days of his concentrated research in Temagami in the 1970s. Hugh is a much-respected veteran of traditional wilderness travel in summer and in winter, and is one of the few people whose encyclopedic knowledge of the traditional travel literature rivals Craig's. Hugh's recollection of meeting Craig is very similar to that of almost anyone else with similar interests encountering Craig. Never one to blow his own horn about his deep and abiding knowledge of traditional place names, travel routes, and travel techniques, Craig more or less just shows up and starts talking. This was the case at the Toronto Sportsmen's Show in 1974. Hugh, a year older than Craig, had just started his Temagami outfitting business, called Headwaters, and had gone to the show to promote his trips. Here's how Hugh recalled their meeting:

> I went for coffee and came back to the booth and there was this guy talking. And he said, "Do you know the name of this lake?" And then he gives us the Ojibway name.

> And then, "Do you know what goes on here?" "Yeah," I said. "That's the five short portages." "No, no," he said. "In Ojibway, it's a word that actually means something else." And I thought, *Who the hell is this guy?* I turned my back on the booth for ten minutes and there's this guy firing all these names in the Native language. Pretty interesting. He must have stayed for an hour. And he came back a couple of times that day. So we exchanged phone numbers.
>
> And then, later on that summer I was guiding a canoe trip in Temagami. And when I returned to Camp Chimo, our basecamp, my wife said, "Remember that guy Craig Macdonald from the Sportsmen's Show who knew all the Ojibway names? He and his wife stopped by while you were gone. It was great. They stayed for two or three days. She did the cooking. She's a farm girl. Not afraid of hard work."[10]

From that moment on, Craig consulted Hugh not only throughout the creation of his map but also through the years of work on his book. It made sense that Craig would ask Hugh — who, in addition to his other fine qualities, was a published author — to write a foreword for his book. Hugh wrote:

> It has often been said that the Canadian climate really only has two seasons — one is winter and the other is "six months of poor sledding." Until Craig Macdonald came along, there were not any serious investigations of what constitutes good sledding. Many of those from earlier times who did know about good sledding lacked the time, inclination or writing skills to record the information. In the mid 1970s on Lake Temagami, my wife, a friend and I were starting winter camping trips to broaden the program of our outdoor adventure company called Headwaters. We knew that there was a traditional Canadian style of

winter travel but were having trouble finding out details and securing the proper equipment. Fortunately, not long after we started with "souped up" Canadian Tire hill sliding toboggans, we were befriended by Craig Macdonald. Our learning curve steepened quickly and our competence followed quickly right behind.

Craig would often drive up to Temagami after a day's work at the Frost Centre in Dorset and, well after dark, head out across Lake Temagami following our trail. Getting a little worried as the evening wore on, we would often drive out on a snowmobile to check on his progress. Invariably he would be moving along with a new toboggan, a new hitch or hauling technique he had just learned about. Whatever the weather, a lively, excited explanation of the finer points would take place, Craig would decline a ride and show up at the camp in a while. Over tea, discussions would continue until the early hours and, at first light, the hands-on explanations commenced. What impressed me was Craig's enthusiasm for the most minute details of winter travel — a passion sustained, unabated, for more than fifty years.

Craig's research was conducted at a crucial time. Written records were minimal. Non-mechanized winter travel was quickly fading from the Canadian woods in the last half of the twentieth century. However, it had not faded from the memory of the Native and Métis Elders and white woodsmen with whom Craig spoke. It takes a special skill and determination to discover who would have this information, locate these people and then gain sufficient trust and confidence for them to share their knowledge. All the fascinating details were gleaned over many conversations and countless cups of tea. Now, it would be impossible to start this project, for, as Craig mentions, all of these Elders have passed on.

> I have long felt that Craig's information on traditional winter travel is of the same calibre and importance as that compiled by Adney and Chapelle in *The Bark Canoes and Skin Boats of North America* (Smithsonian, Washington, 1964). It is exciting finally to have this information out of the boxes and off the pallets in Craig's basement and between two covers. Now, it can be appreciated and utilized by all who value the unique travel styles that evolved from the Canadian landscape.
> Hugh Stewart
> Wakefield, Quebec
> March 2003

Craig's introduction to the book is also illuminating, not only for the depth and range of his research but also for his semi-maniacal approach to detail. This introduction, also written around 2003, gives clues about Craig's utmost humility with respect to the complexity of this knowledge, to the deep wisdom of the Elders who gave it to him, and to why this unique volume remains unpublished to this day. He writes:

> The first version of this book appeared in 1985 as instructions to accompany toboggans and sleighs I had been producing for fellow winter travellers. Soon after the release of these instructions, it became apparent many people hold misconceptions concerning the selection of sleds and toboggans for wilderness snowshoe travel and camping. I rewrote these instructions in 1989 to clarify the basic characteristics of sledding equipment and to provide more detailed information as to their use in the wide range of conditions encountered in North America. For variety in the text, I use the words sled and sleigh interchangeably with no intended difference in meaning.
>
> Ongoing investigations both in the field and through a broad network of Native informants provided more

information for this enlarged and more detailed edition. I've included additional information on strategies for dealing with hills and on the treatment of running surfaces. The sliding process is still far from being fully understood, and research continues to this day. I remain optimistic that new discoveries will aid non-mechanized wilderness travellers in the future.

Unless stated otherwise, my opinions concerning performance and the relative merits of various techniques are based on casual field observations. Since careful measurement and controlled testing are needed to verify some of my observations regarding running surfaces and sliding, consider the information on these topics as tentative and preliminary.

With the advent of the modern snowmobile in the early 1960s, traditional North American sledding equipment was almost completely abandoned for work purposes. However, there is now a growing interest in traditional sledding devices, snowshoe trails and winter camping for recreation. In parks and recreational areas where mechanized equipment is not permitted, traditional North American sledding equipment remains the most practical means for winter travel. It is much easier to sled the heavy loads required to comfortably sustain oneself than to carry them. This fundamental premise holds true everywhere, except in steep terrain.

In the early 1970s, I began making hand-pulled devices for winter camping. These were closely patterned from Native designs used for centuries. Many proved too expensive to build at a consistent standard in any quantity because of the time required to find suitable wood and to bend it into the necessary shapes. In 1978, I first experimented with plastic toboggans and ended regular production of the old "all wood" models in 1980. Plastic

> running surfaces improved sliding performance and significantly reduced production costs. Toboggans could now be made more flexible and sleds stronger and of simpler construction. All my snowshoe expeditions since 1980 have increasingly relied on plastic running surfaces. For nostalgic reasons, I have held on to some of my wood toboggans and leg sleighs but only a few of them have ever touched snow in the past four decades....
>
> May your travels be as hill- and slush-free as possible!

In Hugh's foreword to Craig's book, the comparison of his friend's life and passions to those of Edwin Tappan Adney is entirely apt. The parallels between their lives are uncanny. Born in Athens, Ohio, on July 13, 1868, Adney travelled to Woodstock, New Brunswick, with his family at age eighteen. More or less by chance — as Craig's meeting with Pete Albany at Mowat's Landing on the Montreal River in the 1960s had been — he met and befriended Maliseet canoe builder Peter Jo at Lane's Creek on the shores of the Wolastoq (St. John) River.[11] It was a moment that would shape the course of Adney's life. In his journal, Adney describes the meeting this way:

> When we first came to Upper Woodstock in 1887, Peter Jo or Joseph, on the Point, was living in a house with walls and roof entirely of birch bark, and in summer his old mother did the cooking out of doors, and some of their dishes were neatly made of birch bark also. We built birch bark canoes, tho the Indians now were fastening the gunwales together with nails and no longer with wrappings of split spruce roots in the ancient manner. Even this little which remains of the old manner of living we found fascinating. It seemed to me then as it has ever since from contacts with the Indian in his primitive life untouched by the white man's culture, that the Indian has attained, that which the Japanese possessed *not so much a low standard of living so much as a high standard of*

> *simplicity*, which under the same conditions the white man was not essentially improved upon.[12]

In Craig's introduction to his winter travel manuscript, you can almost hear echoes of Adney's rapture, his respect for traditional life and Indigenous travel technologies, and his appreciation of his informants' "high standard of simplicity." While Craig knew of Adney's dogged research into all things bark canoe, it is difficult to say whether Adney's example inspired Craig to follow in his footsteps as a documentarian of Indigenous travel technologies and techniques. However, the striking similarity between the life passions and products of Edwin Tappan Adney and those of Craig Kennedy Macdonald makes it fair, in my estimation, to call both Craig's map and his book hugely "significant project[s] of cultural preservation in the history of Canada."

Where Craig was trained as a marine biologist, Adney was an accomplished artist by vocation. His avocational interest in the Indigenous languages and traditional ways of Peter Jo and his people throughout New Brunswick eventually drew Adney to settle there and to document the important human knowledge that was disappearing with each passing Elder, each passing day. And similar to Craig's heartfelt effort to convince the court of the value of his work in asserting the historical land use and rights of the Teme-Augama Anishnabai, near the end of his life Adney put on his best suit and appeared before a New Brunswick magistrate, Kenneth MacLauchlan, to argue, also unsuccessfully, that his Maliseet friend Peter Paul had the right to harvest ash saplings on his traditional lands, now technically owned by someone else.

Adney published one book in his lifetime, *The Klondike Stampede*, essentially a compilation of art, photographs, and stories from a trip he took to the Klondike Gold Rush in 1897 and 1898. It was well received in its day. When Adney died in 1950, in his eighty-second year, he left behind maps of Indigenous place names in New Brunswick and northern New England, and an absolute trove of 150 meticulously constructed models, along with voluminous correspondence, notes, and sketches, and an unfinished book manuscript about birchbark canoe styles and building techniques.

Although he had worked as a consultant and correspondent for the Redpath Museum at McGill University, the McCord National Museum, and the Strathcona Museum in Montreal, none of these institutions had any enduring interest in acquiring the proceeds of Adney's life's work. When Adney defaulted on a thousand-dollar loan, the McCord National Museum had taken a goodly number of his models as security on this loan. Happily, they were found in a dusty heap by private collectors Nola and Frederick Hill, who purchased them in 1940 for the value of the loan plus $450 interest for the newly established Mariners' Museum in Newport News, Virginia.[13]

The Hills also convinced the Mariners' Museum to offer Adney, who was always strapped for cash, a stipend of a hundred dollars a month to complete his life's treatise on bark canoes. Sadly, Adney died before finishing his book, but thanks to the Hills, naval architect and boat builder Howard Chapelle was convinced to pick up the project. Chapelle was the curator of the Division of Transportation at the Smithsonian Institution and, like Adney, had made his way with just a high school education. *The Bark Canoes and Skin Boats of North America* was published in 1964.[14] Although the early reaction was muted, the work became a classic and came to be dubbed the bible of bark canoes. On the back cover of a later edition, New Brunswick author and publisher Keith Helmuth lauded it as "one of the most significant projects of cultural preservation in the history of Canada."

Like Adney, Craig was simply too busy with his research to seek recognition or to rest on his laurels in any way. It was as if the process of writing his winter travel opus was, in fact, the destination, for every time he talked to someone new or gleaned another fact or two from a serendipitous source, he would have to add to the manuscript. With Doris as his chief cheerleader, Craig knew he wanted one day to publish this book, but not before it was ready. And as the days, weeks, and years ticked on from 2003, when Hugh Stewart saw the whole manuscript and wrote his foreword, Craig always had energy for revisions and updates but never for actually saying, "Enough is enough," and working to get the book published.

The interviews he did after 2003 were done on a catch-as-catch-can basis with contacts he made through his work in Algonquin Park and through his winter travels with friends and summer travels alone with Doris, since

Craig with Algonquin Elder William Commanda in Algonquin Park. Craig, in OMNR uniform (which he almost never wore), showing respect for the much revered Indigenous knowledge holder.

the kids had all fledged by then. Craig had met Algonquin Elder William Commanda from Kitigan Zibi, north of Ottawa, through Kirk Wipper, but it was only when Craig was assigned by his boss in Algonquin Park to assist amateur archaeologist Bill Allen on a project to document the cultural heritage of the park that he actually sat down to chat with Grandfather Commanda. Bill Allen's acknowledgement of Craig's help with his work indicates his high regard for Craig's stature in the field: "Craig Macdonald was particularly helpful with fieldwork and logistical support. This master of the *nastawgan* has a rare skill at reading landscape. His willingness to share his knowledge, provide references and explain Anishinaabemowin phrases was immeasurably helpful."[15]

Likewise, with the help of OMNR colleagues Mike Buss and Sylvio St. Jules, who set up a rendezvous with the Diamond family who were trapping at the time on the Opasatika River near Hearst, Craig and his colleagues were able head north and set up their canvas wall tent and, for a number of days, trap together with Elders Luke and Gertie Diamond and their forty-year-old son, Victor. The value of these interviews was inestimable because the Diamond family was originally from Waskaganish on southern James Bay and were able to provide access to the perspectives of Cree harvesters in Eastmain, Quebec. As with his conversations with William Commanda and others on Bill Allen's project, his learnings from this time on the land with the Diamond family were profound yet not really consummated until they were written into the manuscript of his book.

The depth of the Indigenous knowledge he encountered in visits like these always left Craig in awe of how much he didn't know and of how much there was yet to learn. Book-as-process seemed a much more fitting characterization of his desire than book-as-product, which necessarily involved a deadline and a point after which no additions, editorial jigs or jogs, deletions, alterations, subtractions, clarifications, adjustments, reductions, addendums, parenthetical embellishments, or word rearrangements for the sake of clarity (or for the sake of word rearranging) could be undertaken, scary as that might be!

The signed maps that Craig had sown throughout the community of Bear Island in 1993 spoke quietly from the walls of houses in which they were hung. With the passing of more and more of the Elders to whom Craig had spoken, the remaining family members realized that life for them was shifting further and further away from the life their forebears had lived — the places they had trapped, the winter and summer routes they had travelled, the place names they instinctively knew, the language written into habits and rituals of living on the land. And so the value of the *Historical Map of Temagami*, with its exclusively Anishinaabemowin toponymy, was appreciating.

Anyone at Bear Island with even a passing interest in other land claim trials throughout Canada — *Calder et al. v. Attorney-General of British Columbia* (1973); *R. v. Sparrow* (1990); *Delgamuukw v. British Columbia*

(1997); *R. v. Marshall* (1999); *Corbiere v. Canada* (1999) — could see that Aboriginal title to land, however constituted by the law, however fairly or unfairly, would always involve testimony about incontrovertible proof of occupancy and use of the land over time. And this proof always involved the special names Indigenous people had for their lands and the stories passed down through the generations of the winter and summer trails — exactly like, and never more exhaustively rendered than, the names and *nastawgan* on Craig Macdonald's map of n'Daki Menan. If the Temagami land claim trial had done what writer Jocelyn Thorpe characterized later as "erasing n'Daki Menan from the Canadian landscape,"[16] Craig Macdonald's map indisputably put Teme-Augama Anishnabai pride of place back out there for all to see.

It was in the spirit of this growing resurgence of community confidence and pride, arising from new optimism that the land claim might finally reach a point of settlement, that Craig was invited back to Bear Island in 2008 to give a presentation about the map. On a winter afternoon the community gathered in the same old rec hall where the academics had met back in the summer of 1983 to prepare for the trial. Craig and Doris had driven up for the day from their home in Dwight. On an easel at the front of the hall was a dry-mounted map, which was the only audiovisual aid in the room. Wearing the same unassuming outfit of jeans, work shirt, and suspenders that had set his informants at ease through decades of research, Craig didn't so much present anything formal as he invited questions that led to stories, explanations, and more stories. Band administrator Victoria Grant, who had convened the meeting, described what happened:

> The whole community was represented, every age, every family, maybe a hundred people in the room, which filled the room. The room was full. And on a Saturday afternoon, if you please. Craig was supposed to speak for forty-five minutes. He went on for over an hour. By the time he was finished he was there for two hours. He tried to break it off, saying, "I'm taking too much of your time," but the questions just kept on coming. Of course — you've dialogued

> with Craig — there are no one-word answers. Every question produces a story, and the stories were mesmerizing.[17]

Next, Victoria stood to close the meeting and thank him for the work that he had done. With that, the room rose spontaneously and began an ovation that continued for several minutes. Craig, excruciatingly uncomfortable, didn't know what to do. Young, old, and in between, the whole community of Bear Island was applauding him, recognizing what he had done, what the Elders had given him, what Doris had supported from Day One. Craig looked at the ceiling. He looked at his feet. He looked at Doris. He looked at the crowd. And still they clapped and cheered. There was nowhere to run, as much as that might have been his first choice of action in a situation like this. Standing in the umbra of the spotlight was the last place he would ever expect to be, or want to be. And yet here he was, at sixty-one years old, receiving the ovation of his life from the people and in the place that mattered most.

Victoria's settler spouse, Dick Grant, who was at the meeting, described it this way: "I think what the community understood was they were valued. This map, this work Craig had done with the Elders, was an expression of their worth and the expression of value was for something they thought they had lost, that they didn't value themselves because of what Vicki's father's and grandfather's generation went through.... [Parents said,] 'Don't speak your language, because being an Indian is a liability in the world, not an asset.'"

Vicki then said, "When he was presenting, it wasn't like he was presenting his information. He made it clear that these were *our* stories. 'These are your stories that were given to me,' he said. I have a picture of him at that event and he has a smile that says, 'Where can I go to hide?' Because they were still clapping. And the expression on his face said it all."

Dick added, "He didn't want it to be about him. All of a sudden it had become about him. If you credit reincarnation, I'm pretty sure Craig was Anishinaabe in a prior life. He had to have been."

Travelling with Craig IV — Rock Star Return to Bear Island

ON THE MONDAY MORNING FOLLOWING HIS RETURN FROM THIS memorable visit to Bear Island in December 2008, Craig got into his truck and drove east via Highways 35 and 60, and then along the Frank MacDougall Parkway to his office at the East Gate of Algonquin Park, as he had done for nearly twenty years. By now, there was no access road, landing, dogsled trail, hiking trail, bike trail, canoe route, ATV speedway, winter campsite, swamp, landing, deer yard, or moose boudoir that he did not know, manage, or have a hand in creating or maintaining. Depending on the season, the weather, the issues of the day, and the priorities of the boss, he might participate in brief discussions with park colleagues in the office about setting work priorities, but more likely he would just make decisions on his own about how best to spend his days in service of the oldest provincial park in the country. Very likely he would have told no one about the ovation he'd received for his life's work outside the OMNR. Not much for small talk, Craig.

The rhythm of the daily back and forth across the park, like a pendulum marking time, had linked birthdays, anniversaries, family gatherings, winter trips, summer vacations, and the comings and goings of Craig's life into one continuous lifeline that wove through and around his enduring passions: Doris, the map, the ongoing research, the book. By 2008, the kids had well fledged and, from the empty nest on Sale Road, Doris had returned to teaching. Their eldest, Nancy, with her master's degree in ecology, was married with three kids and had an excellent scientific communications job with the federal government in Ottawa. Janet had earned a Ph.D. in nanoparticle synthesis at the University of Alberta and was a distinguished professor of chemistry at Vanderbilt University, living with her husband and three kids in Nashville, Tennessee. Colin, who had taken a business degree from Guelph University, began his working life with an unhappy sojourn in bank management. He married and had one child then pivoted, like his dad, back toward his love of the outdoors. He went back to school and thrived, winning the gold medal (top of his class) in the Bachelor of Forestry program at the University of New Brunswick. Like his grandfather, he was now employed by York Region as a municipal forester.

Toward the end of his career in the park, no one really knew where Craig was or what he was doing. Concern that he might get into trouble or hurt himself, or worse, compelled his superiors to put a GPS transceiver on his truck so they would have some idea where he was in the broad expanses of the park, in the event that he mysteriously disappeared. And still Craig went back and forth, back and forth, across the park, until June 30, 2014, when, at sixty-seven years of age, he reluctantly retired from the OMNR after forty-two years of service. While others might have been given a nice plaque or a gold watch to mark the end of a storied career, Craig had no interest in such symbolic gestures. For his retirement, by his request, he was given a chainsaw, a sturdy Swedish Husqvarna, with the understanding by all concerned, as they heaped praise on him at a retirement shindig, that he could use the gift to keep building trails — as they *knew* he would — on his own time.

Craig was approaching the sixth anniversary of his retirement when we began recording conversations for the creation of this book. Besides the ongoing refinement of line drawings for his book on traditional winter sledding, the project of the day, which extended his work in the park both in space and in time, was to reopen and refresh the portages on the Oxtongue River from inside Algonquin Park past his house in Dwight to Lake of Bays. This was the same route, he explained as Doris and I helped to move rocks at one of the portage run-outs for a couple of afternoons (just to break up the long hours of sitting), that explorer David Thompson took on his journey through mid-north Ontario in the summer of 1837.

It was the spring of 2020. Craig was seventy-three years old and we were having to work around pandemic protocols for the indoor portions of our visit. But outdoors, in his classic uniform of suspender-hoisted blue jeans over a forest-green pearl-buttoned work shirt and a Tilley hat, I found myself moving rocks with a new appreciation for Craig-o-matic's capacity for hard physical labour, even if he was nearly ten years my senior. Doris was there, ostensibly to be on hand to pick up the pieces if Craig collapsed, but it was quite clear that the two of them were like a pair of sensible fine leather shoes, as matched and in love as they had ever been — Craig calling the shots and Doris making everything else happen.

On another day, we drove farther up the Oxtongue for a change of pace, right into Algonquin Park, as far as Ragged Falls, where we walked the existing portage to get to the upstream end of the trail. There we stood for a while to take in the falls. We reached the upstream terminus of the portage and turned around, at which point we revisited the story of the great map-maker David Thompson, the guy described by his biographer and editor, Joseph Tyrrell, as "the greatest practical land geographer that the world has produced."[1] Craig began to decant details from his apparently encyclopedic knowledge of not only Thompson's route but also his inclinations and motivations as a trailblazer and map-maker.

Walking slowly back along the trail as it followed the riverbank, Craig pointed to a *V*-shaped draw leading up and off to the left. He said, "Thompson was following the old portage here and there's no way he would have picked along the bank. What you're looking at here is an old treadway that takes us right back to where Thompson would have walked." And with that, we headed up the hill, me with a whole new appreciation for Craig and his close read of David Thompson's contribution to mapping this corner of Canada. Doing my best to keep up, I tripped along behind him, wondering if, somewhere in his heart of hearts, somewhere in his deep and private emotions, Craig *was* David Thompson, in all innocence making his way through uncharted land.

I had similar thoughts when we settled into his office again in the back bedroom at Sale Road. On the wall was a rendering of a New Brunswick map created by William Francis Ganong. Like Tappan Adney, like Craig Macdonald, Ganong — yes, of Canada's oldest chocolatier family — was a passionate naturalist, teacher, author, outdoorsman, linguist, and cartographer, a Renaissance man of the woods.[2] Craig had studied and appreciated his work. He had paddled with daughter Nancy and some of her friends around Ganong's home territory in southeastern New Brunswick. He had weighed what Ganong did and how he did it against his own methods, not out of flattery or adulation, but just because there might have been something in Ganong's work from which he could learn. Was Craig trying in some way to *be* Tappan Adney or William Ganong? I think not. Certainly no more so than he ever thought himself worthy of walking as an equal in David Thompson's moccasins.

It was Craig's humility that continued to astound me when we travelled together to Bear Island in late September 2020 to present the idea of a book to contextualize the life — the lives — from which the *Historical Map of Temagami* was created. Since that heady day in the old rec hall in 2008, the Teme-Augama Anishnabai of Bear Island had gone out of their way to prize Craig and learn from the deep knowledge of their Elders that he uniquely carries. Craig had accepted numerous invitations to speak and to conduct workshops with youth and to travel in n'Daki Menan, each time passing along a little more of what he knows to subsequent generations. In 2018, just before Craig, Doris, and their family were heading into the bush near the Macdonald family cottage on Kashagawigamog Lake in Haliburton County to make maple syrup, Larry Turner's great-nephew John Turner brought a delegation from Bear Island to Dwight, where they made video recordings of conversations with Craig that went on for three days.

Arriving on Bear Island with Craig was a bit like showing up at a party with Rolling Stones frontman Mick Jagger in tow. He's a true rock star, even if he looks more like an old guy who just staggered in from the deep woods, a man who by nature and intent exudes not one iota of celebrity or self-satisfaction. Watching Craig settle quietly into a stacking chair in the Bear Island Canoe House, right on the water where Larry Turner and

Andrew Couchie had embarked in October 1938, brought to mind a distinction made by Janet in describing her dad's character. She said, "In *World of Wonders,* part of the Deptford Trilogy that I read in high school English class, writer Robertson Davies distinguishes between an egotist and an egoist. An *egotist* is so very self-centred that all else doesn't matter. It is plain selfishness. An *egoist,* however, is so absorbed in an idea, a goal, a craft, an ideal, that all else falls to the wayside. While others might see it as selfishness, the egoist sees this approach to life as proper ranking of values and their dedication to self*less*ness. That is my dad to a *T.*"

In addition to Craig and John Turner, in the small circle in the Canoe House were seated Temagami First Nation Chief Shelly Moore-Frappier; Second Chief Johnny McKenzie; Teme-Augama Anishnabai Chief Leanna Farr; Michael Kush Kush Paul's son Michael Paul and his wife, Cathy Metcalfe, Andrew Couchie's daughter; and book initiators Dick and Vicki Grant. After a cordial welcome, Craig began by referring to a copy of the map on the wall. He said, "I'm just a small part of a huge number of people who were involved in making this map. But there is a lot of detail here that can be learned."

And with that he launched into a comprehensive description of the lengths to which he and Alejandro Rabazo went to bring the fall camp image onto the map in living colour. How, in inspecting the glass negative of the original black-and-white picture, they got excellent detail in shading, placement of people and objects, forms, et cetera, but in order to get the colours right they had to obtain a real moose head for making *shizhee.* They had to go to a gun expert to learn that the two guns in the picture were an HBC Imperial #1 cap gun and a specific model of Marlin hunting rifle. He talked about Gary Potts getting the key to the Matchett family cottage on Temeeaygaming, where they photographed the family's nineteenth-century bark canoe to make sure all of the details on the map were authentic. Craig talked about lowering the canoe down from the rafters of the cottage and hoisting it on his shoulders to get the reference shots for Alejandro. "That's my little pot-belly sticking out on the guy carrying the canoe on the map," he said with a laugh.

But this discussion was simply a path into the deeper discussion about the map and its significance to the community. Craig reminded everyone that he had begun this project in 1966, when he was just twenty years old. His great regret was that Chief Alex Paul, who had made the impassioned speech about the impending decay from colonization of Anishinaabe culture to anthropologist Frank Speck in 1913, had passed away the year before. But he had had the privilege, he told them, of interviewing many others, like Michael's father, Kush Kush, who had also since passed away.

At that point, Chief Shelly added, "This map is great insight for us. My father's generation was the last generation to live on the land. People who understood the land like that are all gone. When I was a teen, I wasn't that interested in all this, but now I see the map's value. This for us is so much more that all those Anishinaabe place names. This map tells us what the land was used for."

John Turner, clearly moved by what was happening, said, "This map is the beginning of a new awareness. We need to build curriculum around this."

As this morning discussion threaded deeper and deeper into the map, I was astonished at how things I'd never heard started coming out of Craig's mouth. Talk of medicines, shaking tents, bear walking, nuances of language, quarry sites for ochre and pipestone, burial places, flying skeletons, *weeshna* (beaver testicles), pink lady's slipper and other natural remedies, ceremonies, secret rock painting sites, the portage from Obabika to Wawweeaygaming Cheench Point where unspeakable things had happened, the Wendawban story, and so much more swirled through the Canoe House. Sometimes Craig's stories prompted others in the circle to remember things that brought new stories into the mix, and sometimes the stories from the Bear Islanders triggered another memory in Craig that would spin the conversation in another generative direction. As had happened with every one of his informants, just mentioning the place names triggered or released stories that would be told at no other time. It occurred to me, sitting there in the Canoe House that day, that Dick Grant might have been right about Craig and reincarnation.

Afterwards, over a picnic lunch on the dock outside the Canoe House, I asked Chief Shelly if it was a little bit weird to have a white guy from Toronto teaching place names and telling stories of her ancestors in Anishinaabemowin. "Forgetting is a result of colonization," she replied. "Craig was on our path. We are thankful."

Having seen Craig give many presentations, I knew that the map was a powerful mnemonic release for stories. But I had never witnessed Craig presenting to an all-Anishinaabe audience. By then, he and I had talked for hours and hours on tape, for hours in tents, and for many miles on the trail and on the road, but much of what he talked about in the Canoe House that day was stuff I'd never heard. These were stories given to Craig by the Teme-Augama Anishnabai Elders expressly for the people of Bear Island, and apparently no one else. I was moved by his allegiance to the provenance and sanctity of the story form and yet, while revelling in what Craig had found in his interviews, I ached for what these Elders had taken with them to their graves. Still, I felt privileged to be there and couldn't help thinking of Hugh Stewart's observation that "Craig is like an iceberg. Ninety percent of the information he holds is not visible."[3]

After lunch, we walked up the hill from the Canoe House to the Elders Hall, where there was a regular gathering of folk wanting coffee and companionship and a chance to get out of the house. There were about a dozen people there, in addition to Chief Shelly and Johnny McKenzie, who got things going by opening the floor to anyone who had questions for Craig. As in so many buildings on Bear Island, there was a copy of Craig's map on the wall, but in this case, most of the topics and questions raised didn't refer specifically to the map except in words of appreciation for Craig's role in gathering the information. "We're blessed that the Creator put you in our lives," said one person. "*Chi meegwetch* for all the knowledge you have brought," said another.

Following a lengthy conversation about traditional maple sugar making, where and how it happened throughout n'Daki Menan, Craig captivated the group with stories about how he had taken what he'd learned from the Elders and applied it to his own family traditions of boiling maple syrup in open iron kettles, very similar to the scene depicted in the artwork on the bottom left corner of the map.

Johnny McKenzie picked up on Craig's knowledge of maple sugaring locations and asked, "Are there other places we don't know about that you know about?"

Craig replied, "Maybe. I happen to know that Temagami had pipestone quarries. I've been told where those places are. It's soft black rock. The nearest one is under water right now near the Cross Lake dam, close to the dam, on the left if you're facing the dam. There's a rock bench there where they used to quarry the stone. The stem was wood. Later they used inlays of lead. I interviewed your grandfather Donald McKenzie about all this."[4]

And from there, as with the discussion in the Canoe House, the conversation ranged from *wanumen* (red ochre) deposits on the Yorston River to a cave on the Mattawa River to the chemistry of how the iron in *wanumen* was used as a pigment in rock painting mixtures and on to a woman from McMaster University who did her doctorate on determining the source location of ochre used in particular *mazinaawbikinigin* throughout the boreal.[5] And then, like riding a speeding freight train, Michael Paul talked of visiting sites and gathering ochre. Craig talked of the Devil's Warehouse on Lake Superior. And on the conversation rolled. Somebody who could make rock burn. A story of a trip to Delaware where the teller saw things related to ochre that came from an ancient firepit uncovered in a dig. That discussion careened into talk of the fine birchbark of n'Daki Menan and how it was sought by canoe builders from far afield. And then to the old Indian Day School on Bear Island, the loss of language, and how the parents of the generation to whom Craig spoke had spoken "Indian" at home only when they wanted to keep a secret from their kids. Again, the map on the wall seemed to be the talisman or the keystone for this whole exchange: so different from the here and now, the immediate concerns of the day, but so essentially related to being Anishinaabe on Temeeaygaming.

The conversation had turned to intentionally burning islands right after the snow has gone to promote blueberry growth, when suddenly the door of the Elders Hall burst open and a man in his early twenties bumped into the room. Someone pulled back a chair and gestured for him to join the gathering, but he just stood there, swaying, surveying the room. He eyed the newcomers and started into a diatribe based on the assumption that Craig was some kind of environmentalist. "You got to talk about saving the forest, don't you?" he blurted. "We need to cut the trees. We need the work. We don't need you liberal fucking idiots. Without people working, you ain't got shit."

People in the room were clearly embarrassed, but without missing a beat they ushered the young man out of the hall, urging him to head home, and called the Bear Island Police Service to see if they could help with that. Later, Craig told me that he was pretty sure the visitor was Walter Becker's grandson. The same Walter Becker who, when Craig arrived at the Empire Hotel in North Bay to interview him back in the 1970s, thought Craig might be a representative of the Department of Veterans Affairs, come to investigate irregularities with Walter's pension. "I guess suspicion runs deep in some families," he said with a chuckle.

After the disturbance, the conversation continued for a while on the difference between quartz and chert and how the Anishinaabemowin word for money, *shonia*, actually means "mica," a flaked shiny rock that was used for peering into the future, if you please. As the meeting wound down, Ursula O'Sullivan, Alice Moore's granddaughter and convenor of the Elders' group, remembered the day in 2008 when Craig's presentation so moved the entire community. "Our conversations continued long after you were gone. For weeks we talked about other stories."

Cathy Metcalfe looked at her husband, Michael Paul, and said, "Our culture and our people have been knocked down so hard … by colonization, by residential school, by the law. Shot down for so long." Cathy then spoke of the map and the stories that were being passed around: "Now, this is so important."

I couldn't help thinking, as Cathy talked about her father's journey to Sawgidjeewayawgamawk with Larry Turner on that fateful October day in 1938, of Larry and Andrew setting up their wire antenna and connecting their fancy new radio to the Delco glass batteries just in time to catch *The War of the Worlds*, which provided such an apt metaphor for the ongoing impact of colonization on their lives. The land knowledge Andrew Couchie and his generation held, the traditional lifestyle that took them into a deep and intimate relationship with the land, the sky, the waters, with the stars, the spirits of the four directions, the overworld, and the underworld — all of that had ebbed under the unrelenting settler appetites for timber, minerals, and recreation. All of the aspects of n'Daki Menan that the Teme-Augama Anishnabai had at first freely shared but then lamented when they found them gone or out of reach, behind treaties and other legal barriers they were powerless to breach.

Maybe the hidden superpower in Craig's map was the way in which it charted a course back into the lexicon of the language, and toward confidence, and to a body of knowledge and lost pride that had all but disappeared as a consequence of this very real war of the worlds.

Chief Leanna Farr concluded the meeting in the Elders Hall by looking directly at Craig, nodding to the map, and saying, "When I hear you talk about something my grandpa taught you, I am touched. It gives you a proudness to be who you are. It makes you proud of where you come from."

Craig's face flushed and he dropped his gaze to an empty cup on the table.

Vicki's brother Doug had a road-killed moose he'd butchered in his freezer. So after the meeting we headed across town on foot to fetch the roast he'd offered her. Passing the old Band Office, Vicki talked of her days as the Band Administrator, when she would set up interviews for Craig and he would talk to people in one of the empty offices. She recalled the day when she came into the office and heard Craig's voice and another voice she recognized but couldn't quite place because they were speaking Anishinaabemowin.

This, as it turned out, was her uncle, Sonny Moore, her mom's brother, who was famously reticent when it came to saying anything about anything, particularly if it was a topic that involved any kind of emotional amplitude. And here he was, engaged in a totally animated conversation about n'Daki Menan places and routes with Craig.

Vicki poked her head around the corner into the room where they were sitting. "Since when do you speak Indian?" she quizzed her uncle.

Apparently, Sonny just smiled, his usual reply to questions that don't need answers. "That's what Craig could do," she now said. "He set people at ease and made it easy to talk because he was so interested."

Doug picked out a roast for his sister and, because his home was next to St. Ursula's Catholic Church, where Dick and Vicki had been married and where many of the Elders to whom Craig spoke had been eulogized, Vicki suggested we stop in and have a look. At one time perhaps the most lavish building in the community, the simple frame structure was sided with silver tin accented with red pillars, window frames, and roof. The bell tower and flagpole appeared similarly cared for, though not recently. We made our way around a drainage ditch and a portable bear trap and up the broad steep stairway to the portico, where Doug used a portable drill to remove wooden crossbars that were securing the door.

Inside, simple wooden bench pews were arranged in three rows, with two aisles leading to the Lord's table, which was set with an embroidered linen cloth, candles, and a brass cross, as if waiting for the day when a full-time priest might return to animate the sanctuary. Four tall Gothic arch windows, two on each side, separated similarly shaped plaques that portrayed the stations of the cross. The bright red, blue, green, and yellow glass in the windowpanes that blocked the view of n'Daki Menan stood in contrast and clashed, in a way, with the four colours — yellow, red, black, and white — of a prominent medicine wheel decoration that hung on the simple white-painted wall behind the chancel.

A hooked and fringed wall-hanging depicting Pope John Paul II hung in the vestry as if, in the absence of a parish priest, he was keeping watch over the flock from within. It occurred to me that St. Ursula's might be the only building on Bear Island without a copy of Craig's map on the wall, that being the sacred text of a land-based spirituality and a sacred way of life the Vatican fought — and fights — mightily to supplant. With his weatherproof plastic map case tucked under his arm, Craig sat down halfway back in the pews, saying nothing, while the rest of us wandered and appreciated the trappings of yet another — however beautiful, tranquil, possibly restorative, and well-kept — destabilizing colonial anachronism in the lives of the Teme-Augama Anishnabai.

Before leaving Bear Island, we made arrangements with Michael Paul to meet at Austin Bay the following day to look through the site of the old settlement at the south end of Temeeaygaming. Michael, who, like Craig, was in his early seventies, is the son of Kush Kush and the grandson of legendary Chief Alex Paul. He has been a confidant and correspondent of Craig throughout Craig's research and was one of the few remaining people on Bear Island still intimately connected to the land and the seasons through both his subsistence harvesting of animals, birds, and fish, and his interests in history and in hanging on to the stories of the old ways and the old times in n'Daki Menan.

The next day, Dick and Vicki and I set out in canoes from their home on Deacon's Island to paddle south the five miles or so to the Austin Bay settlement site. I thought of Larry Turner and Andrew Couchie embarking almost exactly eighty-three years before and of how often swirling winds would make canoe travel on Temeeaygaming impossible. Such was the case in 2021. We could have fought our way down the western lee shore of the south arm and crossed over with the southwesterly wind through the islands off the settlement, but that would have been time-consuming and probably altogether too much effort. So we retreated to the dock, traded our canoes for a sturdy Stanley steel motorboat, and cruised south through the waves to catch up with Michael, who had also come by boat.

If there is a place on the *Historical Map of Temagami* that is ground zero for the radiating impacts of colonization for the Teme-Augama Anishnabai, it is this lovely mainland peninsula and secluded western bay between Austin Bay and the south arm of Temeeaygaming. The land was originally surveyed and set aside by the federal government only after it was realized, when Chief Tonené asked for lands for his people to call their own in 1877, that the people of n'Daki Menan had been missed in the Robinson-Huron Treaty of 1850. This was the place to which a number of key Teme-Augama Anishnabai families gravitated as their traditional lives and yearly patterns of movement were disrupted by the presence of the HBC and the introduction of logging, dams, railways, schools, disease, and laws that included the creation of the Temagami Forest Reserve in 1903.

Both Michael's family (the Pauls) and Vicki's family (the McKenzies), along with the Twains and the Whitebears and others, had from time to time set up traditional camps on these well-drained lands at Austin Bay. They arrived here around 1900, and, in addition to continuing traditional harvesting, they cleared the forest to create gardens and room for a small agricultural operation with draft horses, cows, and maybe poultry. The provincial government was never convinced that this was good use of land so rich in red and white pine timber. But when officials got wind of the ravages of the various diseases that were ripping up the Ottawa Valley into n'Daki Menan — first scarlet fever, measles, and scrofula, followed by typhoid, tuberculosis, and smallpox — against which, like all Indigenous inhabitants of Turtle Island, no Teme-Augama Anishnabai had any natural resistance, the province sent a skilled square-log builder, seconded from one of J.R. Booth's nearby logging camps, to Austin Bay to help the families improve their housing.

With Michael Paul and Craig in the lead, we poked through the maple and alder saplings and eventually found the remnants of the lives their forebears lived at Austin Bay, including the remains of a couple of the structures where everybody agreed the McKenzie family had once thrived. Listening to Vicki and Michael and Craig tell stories, my eyes settled on the broad axe marks on one of the visible timbers and the remains of a keenly dovetailed corner joint. Beside it was an ornate iron bed frame. And not far from that, a caved-in root cellar.

"The province eventually decided my grandfather and his family couldn't live here. The timber was too valuable," said Vicki. "So they moved to Bear Island, as others had done, and left this place. In the seventies the Ministry of Natural Resources came in here and burned what was left of these buildings."

What had been a sunny and serene autumn afternoon slowly morphed, as Vicki and Michael picked up cups, plates, and other things their relatives had used when they lived here, into a much more solemn, almost funereal, occasion. Donald McKenzie, Vicki's grandfather, had been one of Craig's key informants. As they had in the Canoe House and at the meeting in the Elders Hall, personal stories — private stories — started to enter the conversation. I took that as a cue to leave the three of them to their reminiscences and to move out of earshot and do some quiet exploring on my own. I saw the remains of some kind of old barge, a newer canvas-covered canoe, a couple of rusty old vehicles, what looked like a sled with steel runners, various bottles, and bits of stoves — shatterings of lives rocked by disease, circumstance, upheavals, and the steady caprice of government. Eventually I just sat down on the shore, overcome with sadness.

Michael had been to this place many times and knew that his family's homestead was on the inside bay, more easily accessed by driving up and around the point to the other side of the peninsula. There were signs of much more recent occupation here, which prompted Michael to tell us the story of his uncle, Leo Paul — nicknamed Mompey — who had passed away five years earlier at eighty-one years of age. Mompey had grown up in this place that was now forbidden ground in the eyes of the government. And when he died, the family at Bear Island loaded up a backhoe, travelled to this site, dug a grave, and buried Mompey's remains. "Right there," said Michael, pointing to a spot in a beautiful clearing overlooking Austin Bay, just steps from where the original homestead had been.

Craig was astounded that they had not thought to mark the grave with a stone or other indicator. "How will anybody know he's buried there?" he asked.

Michael grinned and replied, "His family knows. This land was to be Indian land. But the logging industry thought it was too valuable for Indians, so the land promise was rescinded. But, in a way, Mompey fooled them all because there he lies, in an unmarked grave, for all time!"

We ended the outing by crossing the water just off the Paul homestead site to the opening of a little trail over the peninsula between Austin Bay and the bay to the north that leads to Cross Lake. As we walked along the trail, Michael's and Craig's trained eyes picked out blazes, old cabin footings, and other evidence of cultural occupation of this place and use of this trail, probably going back a hundred years or more. Craig launched into a soliloquy about how these peninsula trails were used to get around bad ice in the winter, all kinds of sneak routes that allowed people to move with the least amount of effort and risk. But after he had said his piece, Michael, always the tease, said, "Craig, I don't think this trail is on your map. You must have missed this one." Craig, unusually for him, didn't have a copy of the map handy, but when he got home, he took a moment to check that it was on the map — and it *was*! — and called Michael to let him know. Of course, Michael knew that all along. Good fun.

On the way north in the boat, Dick and Vicki explained that, although the negotiations began with Chief Tonené asking for reserve lands in 1877 and then lurched on into the twenty-first century, Austin Bay has been included in the lands set aside for exclusive use and control by the band since 1993, when the Temagami First Nation came so very, very close to settling with the government. But what an odyssey it has been for the Deep Water

People to set the bounds on a place they can call their own, a journey that continues to this day, each step a test of patience, resilience, memory, and imagination for the Teme-Augama Anishnabai. And if there is one document that captures some of the pure essence of all that has passed and all of that is yet to come for the people of Temeeaygaming, it is the map that Craig Macdonald helped the Elders transfer from the invisible maps in their heads onto sheets of thick printed paper for all future generations to have and to hold, to see and to cherish.

It was a nasty, cold, rainy day when Craig and I left the Grants' cozy home on Deacon's Island for one last stop on our autumn trip to n'Daki Menan. The Temagami Community Foundation, who commissioned the book about the map, had invited anyone in the area to an informal outdoor coffee-and-doughnuts encounter with Craig. The pandemic raged on, so even though the day was seriously inclement, they had set up a couple of temporary marquee shelters over picnic tables. Craig settled in with a tube of maps and an audience of about a dozen hardy souls who came to hear his presentation. The coffee was lukewarm, the doughnuts were not brand new, the rain was turning to sleet, and still it was as if Craig, in his hoodie and cap that reads "Neen-Gidi-Way-Gamong aka Lake of Bays," drew everyone in and suspended us in the bubble of warmth that was his enduring passion for the map.

Ninety minutes later, after everyone had asked their questions and Craig had provided the fearless, flowing answers for which he is legendary — as if time was of no consequence whatsoever — he opened the fat cardboard tube, pulled out a small roll of maps, and started personally inscribing them to just about everyone at the gathering. These people rolled their maps back up and carried them, like treasure, through the rain to their cars, leaving the snowshoe cartographer still sitting at the picnic table, sipping stone-cold coffee from a paper cup, with just a hint of satisfaction on his wind-burnished face.

Afterword

HAVING FIRST GONE TO TEMAGAMI BY CANOE IN THE SUMMER OF 1967, following the route that Craig had researched for Kirk Wipper the summer before, it seemed fitting to conclude the formal research for this book with a return to Temagami by canoe. I wanted to reconnect with the trails and waterways of n'Daki Menan, to experience them through everything I had learned about the map, and to speak with a couple of key, if elusive, people on my list for whom these lands are home. In 1938, when Larry Turner and Andrew Couchie travelled to Sawgidjeewayawgamawk, the only lines they had to cross that were not the original *nastawgan* of the Teme-Augama Anishnabai were the borders of the Temagami Forest Reserve, which brought with them restrictions on movement and activities like trapping imposed by the provincial government. In 1964, the Forest Reserve was replaced by another provincial timber management initiative called parks.

In fact, by 2021, the original lands of n'Daki Menan were dissected by the borders of nine provincial parks, not including roadside parks or the Skyline Reserve around Temeeaygaming set up in the 1930s with a handshake agreement between the newly formed Temagami Lakes Association members and the Temagami sawmills and forestry companies. The anchor in this network is Lady Evelyn-Smoothwater Provincial Park, a block of 72,400 hectares of land stretching westward from Mōnskawnawning Sawgihaygunning to Shōnjawawgaming.

The other eight parks are corridor lands that attempt to keep the loggers and the wilderness canoeists happy. By now, traditional use of these lands has disappeared altogether from the provincial maps. Makobe-Grays River Provincial Park protects lands on either side of the Makobe River and connects Lady Evelyn-Smoothwater to Elk Lake to the north. Obabika River and Sturgeon River corridor parks flow southward from the highlands of Lady Evelyn-Smoothwater in a *Y* configuration reaching well beyond the southern limit of Temeeaygaming. The theory behind the corridor park is that a wilderness paddler might hear the logging equipment over the hill, but the logging would not be in plain view of the rivers, thereby protecting the working loggers from having to watch people recreating. The other parks in the network protect other corridors and ecologically significant spots, including the spectacular shores of Sawgidjeewayawgamawk, now contained within Solace Provincial Park. Job one for getting a Temagami canoe trip organized was to get my park permits sorted out.

Since my scramble up to the fire tower on Maple Mountain when I was eleven, I had come to learn through Craig's work that if there is a spiritual centre in n'Daki Menan, it is Cheebayjing — the Place Where Spirits Dwell — the high point on the spectacular rocky spine east of the Ishpatina Ridge. It seemed fitting that this should be one stop on my July 2021 itinerary. I also had an invitation to visit outfitter, conservationist, and author Hap Wilson at his ecolodge at Cabin Falls, over on the South Branch of the Lady Evelyn River, to hear his take on Craig's work. The other two destinations on Craig's map I was able to work into the route for this solo jaunt were Teme-Augama Anishnabai Elder Alex Mathias's place on Obawbika Sawgihaygunning and a final stop to chat with my old friend Hugh Stewart at his place at Camp Temagami on Wigwasati Island in Temeeaygaming. Time would not allow me to travel in a circle from Bear Island, as the lads had done in 1938, so I opted to hop by floatplane to a lake near Cheebayjing and meander my way back to town by canoe, up and down the legendarily rugged portages of what paddlers call the Golden Staircase, leaving plenty of time to visit, soak in the scenery, and ponder the project with rhythmic meditations of movement along ancient paths and paddle lines.

There was nothing rhythmic or particularly meditative about the two-and-a-half-mile hike up Cheebayjing. Strangely, my eleven-year-old self retained no memory at all of exertion on this steep trail. The climb in 2021 was a different story. Stopping often to wheeze and catch my breath, fifty-four years later, my slightly paunchy sixty-six-year-old self asked, "What on *earth* were you thinking? You sit on the couch for a year and a half, isolating for the pandemic, eating Cheezies, drinking beer, and then somehow you think you can put on a pack and jog up one of the highest peaks in Ontario? Think again, old-timer!" Funny how voices rattle through the head of a solo traveller.

Where the climb in 1967 was more about the physicality and camaraderie of scrambling up the metal rungs of the tower and looking out over the route we'd paddled from Lady Evelyn Lake, this time, mercifully, the bottom few rungs of the ladder had been removed and the trap door on the cupola appeared to be locked, so there was no temptation to climb the tower once again, even for old time's sake. On this July day, however, with me were the stories and layers of meaning emanating from Craig's map, of the vision quests that had happened here, of the spirits who dwell on Cheebayjing. Looking south, absorbing the wild beauty and enormity of n'Daki Menan, I cut chunks of cheese and sausage and watched a bald eagle soaring high in the blueness of a summer sky. Closer in, a pair of ravens chatted and cavorted in the updrafts created by the mountain. This was no ordinary place. This was Cheebayjing and I was an interloper here, even with my expensive park permit.

In swept a memory of how incensed I was during my last year of high school, having had the experience here in 1967, to learn that the provincial government was looking for ways to boost the economy of northern Ontario and that what they really wanted to do was turn Maple Mountain into a fancy ski resort — Teme-Augama Anishnabai be damned. So it was with a certain amount of zeal that I supported the efforts of paddling friends in Temagami who had founded the Save Maple Mountain Committee to create a branch office of the organization out of my residence room in Morris Hall when I headed off to Queen's University in the fall of 1973. Happily — and this may be the only good thing that came out of the protracted legal process surrounding the Teme-Augama land claim — when Chief Gary Potts and lawyer Bruce Clark applied to have a land caution put onto the whole of n'Daki Menan that same year, Maple Mountain was spared from development with no further work necessary on the part of non-Indigenous do-gooders. Now that Cheebayjing is part of Lady Evelyn-Smoothwater

Provincial Park, it looks like that is how it will remain for time immemorial, or at least until another government decides otherwise.

Sitting there in the sunshine, my mind casting through other stories of this place, I could hear Elder Mary Laronde's hearty laugh. We had spoken at her home in North Bay about Craig and his work, and about her attachment to this place as a Teme-Augama Anishnabai Kwe who grew up on Bear Island. She had heard about Cheebayjing as a young girl, but by then no one really came here anymore. The place names were fading. The *nastawgan* were growing over.

And yet, as a communications officer for the band, Mary had attended every single day of the land claim trial in that oppressive settlers' sanctum in the Superior Court of Justice in Toronto. She had heard the names, nearly five hundred of them, spoken out loud and described by Craig to the judge. Even in the failure of that case, the trial was a turning point for many people, including Mary. One of the things she did, as many others did too, was to come as an adult to Cheebayjing. And it was here, in this spot, looking out over her homeland, that she reclaimed something, an important insight that fanned embers of deep pride. Here's how she described what happened:

> When I was a little girl, and I was in school, I was taught that unless you created the wheel, you didn't go anywhere as a society. This whole idea about the wheel and people were able to move. And then all of a sudden we're into the industrial revolution and everything is just wonderful. That whole thing about *Breastplate and Buckskin*, that Grade 5 history book. And the Grade 8 one said Pontiac led a rebellion in 1763. I have a copy of that somewhere, because it's so devoid of anything that's real, from my point of view.
>
> I had this feeling that we were just dumb Indians. We didn't even have the wheel. So that's what I was taught. There were certain things that were marks of this great intelligence, this big breakthrough or whatever. So I'm standing on top of Maple Mountain, that sort of south,

> southwest view — I might have been about thirty-seven at the time — we've paddled there.
>
> So up there. I'm looking out over. This is just my education. I'm going, Holy shit! No *wonder* we didn't have a wheel. It would have been completely useless here. We made what we needed. Canoes and snowshoes! That was like a big epiphany for me. Because I was indoctrinated with all that shite they stream on Acorn TV.[1]

That July evening, I sat with a hot cup of tea on a gorgeous rocky campsite on the south shore of Hobart Lake, where we had also camped in 1967, looking north to the sacred mountain, my body still humming from the exertions of the day, my heart full of the complexities of the place, yet feeling strangely satisfied and at peace to have had the honour to be here and to have had the opportunity to thank Cheebayjing for the lessons and learnings of the experience.

How upsetting it must have been to welcome strangers to this land, strangers who tread your trails, cut your trees, deny your language, remove your children, Christianize your ceremonies, trap out your marshes, bring disease, supplant your laws, rename your sacred places, and then draw lines and set boundaries that prohibit access to the places where children yet to be born might commune with the ashes of their ancestors. How gratifying it must have been, in some small but significant way, when the courts were doing their level best to erase your presence on the land, to have a man come along who learns your language, listens intently to your stories, and then slowly, methodically, gently, scribes the routes and the names and the waterlines on a map that shouts "You were here! You *are* here! You will be here as long as you love the land!"

That night, I slept the sleep of the dead. And in the morning, as the rising sun caught the tip of Cheebayjing, a story from June Twain, Bear Island's oldest resident, was rumbling through my aching bones. Like Mary, a visit to Cheebayjing as an adult helped June affirm the process of emancipation from the strictures of the Catholic Church and re-embrace the nourishing land-based spirituality of her forebears. Here's June's story:

I didn't grow up with the ceremonies. We were all very strict Catholics. My grandmother was a very strict Catholic, so I didn't have that when I was growing up. But I went to take a drug and alcohol counselling program at Canadore College. Native spirituality was a strong concept of the program. So that's how I picked up the spirituality. And once I picked up that spirituality, I wanted to learn more. So then, I would go to ceremonies and circles, when we would pass around an eagle feather. At the friendship centre. And then my husband became involved too.

After I finished my course, I got a job in the community as a drug and alcohol worker. I told Jim,[2] my husband, "Let's go up to Maple Mountain." Gary Potts used to talk about Maple Mountain using its Native name, Cheebayjing, saying, "That's where your spirit goes when you pass."

I said, "Let's go up there and have a ceremony up there. We'll take some of the youth." I think we took eight youth and four chaperones. We had no idea where we were going. We'd never been out there before. We went to Diamond Lake portage there, took canoes over the portage, then we paddled from there to Maple Mountain. It was a week-long trip. Beautiful weather. I think it was in August when we went up there.

So we got to the base of the mountain. We thought we'd be able to camp there. But it was all swamp. We were paddling back over to the one campsite on Tupper Lake there, and we were paddling back and suddenly this hawk came down, whoosh. [June laughs.] It was kind of calm and its wings were going. He went and did that four times. Four times he did that, you know. You could just see the ripples in the water.

Some of the people were getting kind of freaked out. The adults asked, "What's that mean, June, what's that mean?"

I said, "I don't know what it means." But I had a feeling that the hawk was coming down to welcome us, telling us … making a clapping sound with his wings as if he was applauding us. He has come to help us pray.

[June tears up and starts to cry.]

Every time I talk about this, I get so emotional. That bird was telling us, "This is yours. This is yours," like, you know. To have an animal kind of participate in telling us, saying, "This land, this mountain is yours."

We went to the top of the mountain the next day and I thought we should do a smudge. Jim had never done a ceremony before but he made a little fire. We gave the kids tobacco and explained to them what tobacco means. So they kinda giggled because this was all new to them too. Jim had to wait for a while to do the actual ceremony because there were tourists up there — there are *always* tourists up there, right, because it's such a beautiful place.

After the tourists left, Jim said, "Okay let's have a ceremony." He told the kids to hold tobacco in your left hand. "That's the hand that's closest to [your] heart," he said. "And then you pray and whatever prayers you're asking for you put them in the fire. And then the prayers go up to the Creator."

We were almost finished the ceremony and in comes this hawk in the eastern door. It was so close we could reach out and touch it. Oh my god! The kids were like this, laughing in awe, this bird coming to join us at our ceremony.

There was two girls on their moon time and they weren't allowed to be a part of the ceremony. So the hawk went and circled around them too. And then he left.

That kinda clinched it for me. "This is mine. This belongs to me." I don't know what other people got out of it but I certainly got a lot out of it. It changed my life, really. My spirituality.

> The next day, we were leaving. I think we had eight canoes. They all went out ahead. Jim and I stayed behind to make sure that the campsite was clean and the fire was out. We were the last ones to paddle out and the hawk came and circled around us again.
>
> Before that, Jim and I used to go to [the Anishinabe Spiritual Centre run by the Jesuits at] Anderson Lake, where they were teaching people to become more involved with the Church. Jim was trained to be a deacon. I was training to be a minister of the Eucharist. We went down there for four winters and we were getting deeper and deeper into our Catholic faith. But that trip to Cheebayjing changed all that. That trip to Cheebayjing to me is the most profound thing that has ever happened to me spiritually.[3]

June Twain, in her work with Bear Island youth, was among those who most clearly identified the potential of the *Historical Map of Temagami* for curricular purposes. She said,

> When they pick up the language more — and the language is coming back in the community — I think as they pick up the language more they'll probably use the map as a guideline too for the language, to see … this is what this means. Right now, a lot of the youth and a lot of people are travelling out on the land. They're spending a lot more time on that land than they did like, say, twenty years ago. So they're going out there doing ceremonies, and they're doing a lot of things that I didn't grow up with. The map will help with all of that.

Likewise, the map illuminated my continuing journey from Hobart Lake toward Majamaygoz Zibi, the Lady Evelyn River. Crossing from the creek leading to Kawkinōzhēnseswaw, where the water was so low I had to drag the canoe through the muck, I used a waterway that was only created when the logging dams raised the level of Mōnskawnawning, now Lady Evelyn Lake. Making that course adjustment, I thought of Wendawban's cabin floating away, of his potato garden on the island where I had first stayed on that lake, and of stories of his shaking tent and his conjuring that still swirl in the air throughout n'Daki Menan. A young black bear greeted me at the first portage to enter the Golden Staircase and wandered on as I shouldered my gear to start heading upstream.

This treasured moment with this handsome bear reminded me of yet another weird artifact of the cultural collision between the Teme-Augama Anishnabai and those who came from away. According to anthropologist Frank Speck, who spoke to Second Chief Aleck Paul in 2013, the original name for the big island in Temeeaygaming was Makōminising, meaning "Berry Island," named for the blueberries that still grow in profusion in the thin acidic soils of the cracks and hollows in the ancient rocks of the Canadian Shield. Knowing the language better than anthropologist Frank Speck did when he spoke to Second Chief Aleck Paul back in 2013, Craig doesn't believe this alternative derivation of the name. He suspects that Speck somehow mixed up the idea that Bear Island always had lots of blueberries in summer that attracted bears and that this linguistic nuance was lost in translation. Somehow, because *miin* (Anishinaabemowin for "berry") and *makwa* (Anishinaabemowin for "bear") sound alike, Berry Island became Bear Island, a name that has stuck, for better or worse, to this day. In the meditation of my considerable exertions to carry three loads to the head of this first portage, I wondered if this was like having a conversation with an all-knowing correspondent who insists on calling you by a name other than your own. Annoying.

Similar to how the trail up Cheebayjing seemed steeper and more forbidding in 2021 than it did in 1967, the portages up and down the Golden Staircase presented distinct challenges for a senior citizen. But what a delight were the hidden respites those efforts revealed, and what a balm were those nights alone at rocky campsites under the circling stars, with the rush and roar of whitewater to lull me to sleep. But I'd be lying if I claimed that there wasn't certain pleasure in a sauna, a swim, a hot meal, a cold beer, and the company when I arrived at Hap Wilson's place at Cabin Falls, on the South Branch of Majamaygoz Zibi.

During more or less the same period in which Craig has been doing his research, starting in the 1960s and continuing to this day, Hap has been a champion for canoeing, conservation, and wilderness travel throughout n'Daki Menan. Hap worked for the OMNR from 1969 to 1976, and he explained the different work they did this way: "Craig's directive was in the preservation and location of Aboriginal place names. My directive was to define, consolidate, and preserve the canoe routes from the encroachment of logging, and to 'spread out' the user traffic of canoeists so that the most popular routes would not get overused."[4] Hap was an artist and map-maker in his own right, and his illustrated book of Temagami canoe routes, first published by the OMNR in 1977–78, remains a key reference for paddlers. Interestingly, however, after our dinner, the author of that dog-eared key reference book reached into a shelf and pulled out an equally trail-weary copy of the *Historical Map of Temagami*, which he uses to help guests at the Ecolodge appreciate the scope of the historic use of the *nastawgan*, now serving paddlers more than trappers or families from Bear Island.

When his work with the OMNR ended, Hap started an outfitting company that morphed into what he has been undertaking on the site of the old Rapidan Club on the South Branch of the Lady Evelyn River. He co-founded the Temagami Wilderness Society (now Earthroots) in 1986 and, among other spirited anti-logging activities, joined Kush Kush, Gary Potts, and a host of other committed Teme-Augama Anishnabai in their blockades of the Red Squirrel Road in 1988–89. Perhaps more than anyone, Hap suspected that there were political energies powering the delays in the publication of Craig's map: "I remember that we pleaded with Craig to get his map published because it would have helped in our battle to protect the *nastawgan*. But Craig was a 'Company Man,' and, sadly, his hands were tied. The OMNR wouldn't allow the map to be released until it was to the Ministry's benefit and not an aid to protectionism."[5]

When I asked about his connection to Craig, Hap — who's a couple of years younger than Craig — said that he had only met Craig once or twice in passing but knew him by reputation as much as from personal experience as a true wilderness man. Hap put him in a league with some of his other heroes like Grey Owl, John Hornby, Ernest Thompson Seton, and Calvin Rutstrum. Still, it surprised me when he said, "You look at guys like Craig. When he's gone ..." and then had to pause because of the upwelling of emotion. "When he's gone, that whole tradition is gone. It's tough ..." Half a minute of silence ticked by, the rush of Cabin Falls floating in to fill the void. Somewhat recomposed, he continued:

> I don't like the canoeing scene now. Kinda wears a hole in my heart. People coming out here for all the wrong reasons. I think Craig realizes that too. And I think that's one reason that drives him to keep these traditional travel traditions alive. I think these little things, like traditional fire irons and prospector tents, these define part of Canadian history that was more pure, more connected to wilderness and to nature.
>
> I think many people have distanced themselves from what's really important. That's what we're trying to do here at the Ecolodge. We're trying to reconnect people to what's important. Yeah, so talk of Craig's work hits a sensitive part of my life, when I see people's attitudes changing over the years. We're still losing wild spaces after so many years trying to protect it. I certainly don't want to be known as an apocalypticist or fatalist, when it comes to the environment. But I don't have a lot of faith in the system the way it is now.[6]

That Craig, an Ontario government employee, would earn the admiration of hardcore environmentalists and the leadership of a First Nation striving to reclaim its hold on the same land seems somehow contradictory, but it speaks to the breadth, significance, impact, and authenticity of his work.

Getting ready to continue on from Cabin Falls, at Hap's invitation I climbed another long steep trail to a lookout high over the river where a stray cell signal from somewhere in the Ottawa Valley (said Hap) innervated my phone, liberated from its waterproof case. Among the *bing-bing-bing* of a week's welter of texts and emails came news that my Inuit friend Mary Simon had been named the new governor general of Canada. I couldn't help wondering, as I sent a congratulatory note, if Craig's half century of slow and steady valuing of Indigenous ways and traditional knowledge played some kind of role in setting the stage for this head-turning, if overdue, appointment. So ironic that the namesake of Lady Evelyn Lake, a n'Daki Menan settler name that Craig supplanted with the Anishinaabemowin original — Mōnskawnawning — was the daughter of the fifth Marquess of Lansdowne, who was governor general from 1883 to 1888, and the wife of the Duke of Devonshire, the governor general from 1916 to 1921. Such confusion from changing times notwithstanding — an Indigenous woman becoming the Queen's representative in Canada ... really? — as I walked back down the hill amongst the majestic pines, I decided that if I got my wish, it would be Her Excellency Mary May Simon who would present Craig with the Order of Canada for creating what may be the country's most important historical map before her term was done.

Very much swayed by the fineries of "camping" at Cabin Falls Ecolodge and seriously concerned that I was pushing my luck with the physical challenges of the portages to get here without injury — hard as that was for an aging adventurer — I took advantage of a flight that had brought new guests to Hap's place to hop to calmer waters below the Golden Staircase. Next stop on this journey of discovery was the home of Peedankun — meaning "Coming of the Clouds" — also known as Alex Mathias, who has been called everything from a Holy Man and Mystic to New Age Indian and the "last Temagami Aboriginal living off the reserve on ancestral family land."[7]

Before connecting with Alex, I set up camp on a spectacular site at the north end of Obawbika Sawgihaygunning, threw some tobacco and a water bottle into a day pack, and dawdled north through what many consider to be the largest continuous stand of old-growth red and white pines remaining in the world. As I had come to learn, as with other places of spiritual power and significance Craig learned about through his interviews, there is no clue of the importance, energy, and intensity of this place on Craig's map, but he did make a point of insisting I should visit and pay my respects when travelling through. "Check out Cheeskonabikōng, meaning 'At the place of the huge rock,'" he said.

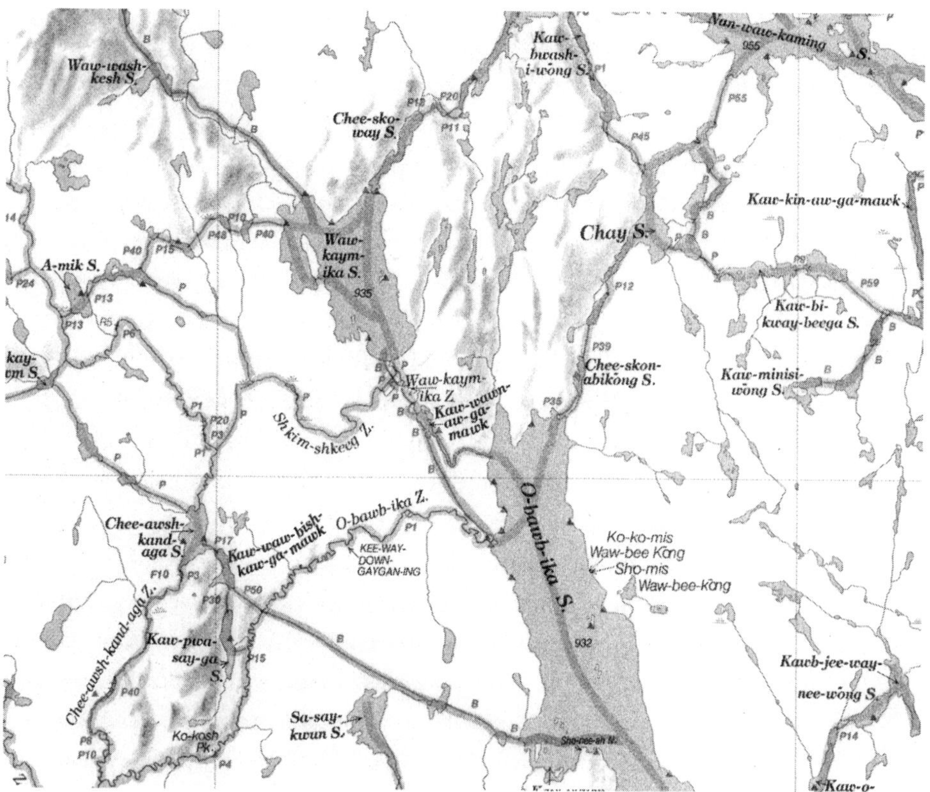

Spirit, Prayer, or Conjuring Rock, as this place is sometimes called, is at the centre of the Wakimika Triangle, which is protected within Obabika River Provincial Park. Although Craig's map betrays nothing of the importance of this place, it didn't take much digging to learn that this ancient sacred site is where Teme-Augama Anishnabai and many other First Nations people have come to fast, to heal, and to dream, or that geomancers and spiritualists from all over the world have found their way here for one reason or another. It's difficult to voice what it felt like to come down to Cheeskonabikōng Lake and behold these two great columns of rock and the significance they radiate. I stripped and swam to rinse off the sweat of the day, made an offering of tobacco, and sat until it was nearly too dark to see, before ambling "home" to my tent on Obabika. I fell asleep listening to what I took to be Peedankun's dogs barking a few miles across the lake.

The following day, heralded by much splashing and crashing at the end of the portage from Cheeskonabikōng to Obabika, just a few hundred yards from where I was having my breakfast, three red canoes were flipped from the shoulders of fit young camp counsellors onto the beach. Over the next couple of hours, so much slower on the trail than their leaders, teenage campers appeared one by one at the end of the trail. Their packs were big, but not of size sufficient to merit the amount of moaning the kids were doing about carrying them. This was a canoe trip of teens from Langskib, a summer tripping camp based on the north end of Temeeaygaming, which, like a couple of dozen other Temagami canoe-tripping camps, have been helping young people to "grow into the best versions of themselves [and learn] how to be with the land, the water, the sky, how to be with one another and ultimately how to be with themselves."[8]

Chatting with the staff as the would-be adventurers straggled in, I learned that they were out for a couple of weeks, more or less just travelling from lake to lake on the old *nastawgan*, the journey always trumping the triumph of arrival at the destination. The self-assured lead guide, with a ponytail and a feather in his fedora, reminded me of a guy a lot like me when I had the world by the tail and knew what was what in leading wilderness canoe trips.

"Did you stop at Cheeskonabikōng?" I asked.

"Yes," he replied. "We did a little ceremony and each of the kids made an offering of tobacco."

How heartwarming it was to look at those grubby adolescent hands and faces belonging to kids who were almost exactly the age I was when I first came through here as a Kandalore camper, and to know that what they were finding to be wicked *hard* in the moment would be wicked *good* in the fullness of time. And how cool it was to know that Indian nights at summer camp had, mercifully, morphed and evolved into culturally appropriate gestures, like prayers to the land and offerings of tobacco, to recognize the original inhabitants of n'Daki Menan.

With Hap's voice and the image of these young canoe trippers very much front of mind, it was clear to me that the overlap through time of traditional families, trappers, loggers, cottagers, campers, and canoe trippers in these sovereign lands of the Teme-Augama Anishnabai probably had a lot to do with the perpetuation and care of the *nastawgan* of n'Daki Menan. In other parts of the country, as documented by the Dene Mapping Project, the Inuit Land Use and Occupancy Project, and other efforts to capture culturally significant information, the trails, the sites, the evidence of human occupancy are separated in vast areas of land — few people, massive places. Here in Temagami, however, the scale is smaller, more intimate, and somehow the Teme-Augama Anishnabai, the cottagers, the loggers, the parks people, the conservationists and environmentalists, and the paddlers have all managed to coexist.

With all of that running through my mind as I brewed another coffee and savoured the intoxicating beauty of Obabika, I thought of Andy Buckman. He talked about a "historical coincidence" that had happened here in the context of Craig's work. When I asked if he'd seen or heard of anything quite like the *Historical Map of Temagami* in his travels, he replied,

> In my experience, Craig's work is unique. It just so swirls above any kind of formal ethnography that I've ever seen, or cultural interpretation. I can't think of anything that compares to it. Of course, he did have incredibly rich material to work with, in an area that, as you know, as the result of a number of sort of historical coincidences, still had much of its traditional trail system … in existence, due to the canoe trip camps and the history of canoe tripping and tourism. And also still had an existing community of Elders who had grown up in the bush. There aren't that many places that have that kind of dense concentration of resources. But Craig made the most of it. He realized … long before anybody other than the Elders did, the depth and the richness of knowledge about that particular piece of geography.[9]

Alex Mathias wasn't there when I paddled across to connect with him.

"He's out on the lake showing some friends around," his daughter told me. "That's probably him there," she said, pointing to a tiny buzzing point on the water a long way down the lake.

I explained that her dad and I had spoken on the phone and that I was camping at the head of the lake. If he didn't happen by the big campsite with his friends, I'd come back the following day. Either way, it would be good to talk to him. And could she let him know that the writer-guy he'd talked to on the phone had dropped in?

Halfway through the paddle back to the big campsite, the sound of an outboard filtered in on the wind, and the next thing I knew I was holding the gunwale of Alex's tin boat, being introduced to his friends and passing the time of day, as if this was just what happened when you were in somebody's home neighbourhood. Alex explained that he had things to do the following day but could come by the campsite after supper to chat.

"Would you like to come for supper?" I asked.

"Do you have fresh fish, beaver, or moose meat?" he replied.

"Naw," I said. "Just freeze-dried horseshit."

"No thanks. I'll have supper at home and come by after that," he replied with a grin.

Good to his word, Alex, seventy-three, turned up as the shadows of pines and thin birches were starting to lengthen on the big campsite. He began by telling me how proud he was to have the map hanging on his wall across the lake, and how he had to explain to people that Craig's handwriting on the bottom, which says, "Thanks Alex for helping with the map," refers not to him but to his adopted father, Alex Misabi — also known as Shimmy — who was one of Craig's informants, also now gone. "I've got to correct people that it wasn't given to me. But it's a great thing to have."

If there is a keeper of the Wakimika Triangle and all of its secrets, it would be Alex Mathias. In 1965, Shimmy decided that life would be easier for the family on Bear Island, so they moved. Alex told me he remembered how sad he was when they walked over the portage to Temeeaygaming for the last time. "But," he said, "I knew I would return." And that is exactly what happened. Fed up with the hard-drinking life of a contractor and hammer-for-hire on Bear Island; tired of the poverty, bitterness, indignities, violence, and crime in families wracked still by colonization, he joined the Temagami Wilderness Society's blockade of the Red Squirrel Road in September 1989. After the disappointments of the trial and the catastrophic

Alex Misabi, a.k.a. "Shimmy," on Obabika River.

schism in the ranks of the Teme-Augama Anishnabai that blew up in 1993 over the land claim settlement, he moved back to the mouth of the Obabika River, where he has been ever since, living with his partner, Mary Carol Hill, who passed, and now with his daughter Natasha and her partner Brandon.

He pointed to the simply rigged long buffet table in the middle of the campsite and explained that since he'd moved back to the land, he had been hosting a gathering in the Wakimika Triangle on the weekend after Thanksgiving each year for anyone who wanted to join in. This was not just for Indigenous participants. Anyone was welcome to attend. He explained what happens this way:

> Cheeskongabikon is a sacred site for our people for thousands of years. And it's still sacred. You go there. You can get the vibe. You get good vibrations. I have a lot of people come here, not only Native people, but also non-Native people. They come here just to get energized and go back to the city, wherever they're from, and they have plans the next year to come and do the same thing. So it's a great, powerful place.
>
> And also there's old growth in the area. Big trees that ... a forest that has not been disturbed yet by man. But it's such a small area. Even though they say it's largest old growth of red and white pine in the world but it's only about three square miles left of it. When I think about what we had throughout our whole lands, how it must have looked at one time ... because just in the last thirty years, I started hearing the term "old growth." I'm sure old growth existed all through here but nobody ever mentioned it before.
>
> People that come here have to understand what the land really means to the people who live on it. Even though it was passed to me, I never look at myself as an owner of the land. Nobody should own the land. I believe I'm just passing through. I'm just a caretaker of the land while I'm alive. It'll be passed down to someone else, when I move on.[10]

It surprised me to hear that, while Alex had not given up on his own people to protect, defend, and validate the remaining treasures of n'Daki Menan, he had actively reached out to non-Indigenous allies to do just that. Perhaps inspired by Hap Wilson and the people who created the Temagami Wilderness Society, Alex joined and remains a board member for another environmental non-governmental organization called the Friends of Temagami, dedicated, among other projects, to keeping the Wakimika Triangle as it was for all future generations, whoever they may be, wherever they may come from.

As the light of our fire slowly replaced the last rays of the setting sun, I asked what it was like living in a park, with gates and controlled access, and whether that had cramped his style at all since he moved back onto the traditional Misabi family territory that is no longer — if it ever was — recognized by the Crown.

"Fine," he said with a big smile. "They locked me in. They locked me out. I have a gun so I just shot the locks off. Simple. Eventually, they gave me a key."

Before heading south and east to retrace Larry Turner and Andrew Couchie's steps through the rocks of the Abikon Abikush portage to Waskiggamawsing and to catch up with Hugh Stewart at Camp Temagami, after the wonderful intensity of the interactions at the big campsite, I wanted to have a little more time alone to regain the rhythm and purpose of the bigger journey by exploring the storied petroglyphs, another site that Craig had purposely decided not to include on his map, on Wawkaymika Sawgihaygunning. So I packed up a few things and made a day trip north, up the Wawkaymika Zibi to Wawkaymika Sawgihaygunning.

The day on Wawkaymika was glorious, the lake millpond smooth, the pressure high and steady, the sun relentless. By the time I settled my canoe onto the smooth rocks of the tiny island cluster off the eastern shore, after slowly circling to see all that there might be to see, it was so hot that the only natural thing to do was strip and cool off in the lake. One of the great pleasures of dips like this is the immediate release of heat into the cool clear water, replaced by the sensation of energy flowing from the centre of the earth into every organ in your body.

Totally invigorated, I clambered out onto the bare rock on the open lakeside of the biggest island in the cluster and spent the next couple of hours running my fingers over the shaped depressions carved in the rock so many years ago. There are human forms. A moose. Another smaller image that might be a bird — a goose? While pictographs, painted on rock with some mixture of ochre and fish oil, are relatively common and of much more recent origin, these physical images, which are chipped into the granite by hand, are a tangible human connection to much older times, thousands of years, maybe more.

Suddenly, distant laughing voices skittered my way over still water that mirrored the sky. Three canoes, heading south on the regular canoe route from Cheeskoway and Nanwawkaming, had rounded the point and appeared to be coming directly toward the islands. After scrambling to put my shorts back on, lest I be confused with Mowgli in some Temagami variant of Kipling's *Jungle Book*, I watched their paddles flash in the sun. The paddlers were chatting away, chatting away, getting closer and closer. My canoe on the shore was likely visible to them, if they were looking, but I was hoping beyond hope that they wouldn't stop here and snap my little revery. Happily, they passed on the inside of the islands and continued south, obviously in a bit of a revery of their own in the Smoothwater Wilderness.

I stretched out on the sun-warmed rock and dozed off. Sometime later — I had no idea how long I'd been asleep — I awakened to a startled croak. Turning slowly onto my side to look in the direction of the sound, I saw a sandhill crane step up on a little ridge of rock not ten paces away. It just stood there, looking in my direction. I had seen a pair of these elegant storky birds feeding on exposed lake-bottom mud as I was crossing over through the flooded lands near Wendawban's place a week or so before, but I had taken no real notice, beyond remembering the first time I'd seen these magnificent birds on a buggy day on the Arctic tundra. But this time, I had a profound sensation that this bird had something to say. I sat up. And still it stood.

For perhaps ten minutes I looked at the bird and the bird looked back at me. Then, with no fluster or fear, it turned and started walking in its stilt-like way in the other direction, looking over its grey-feathered shoulder at one point as if to say, "Come on, I've got something to show you." I got up and, in my bare feet, picked my way along what now seemed hallowed ground — perhaps as vision questers had done many times since those images were carved in the rock.

In my head, from a cluster of emotions, I tried to tell the bird that it had nothing to fear from me. But it seemed already to know that. I followed as my guide picked their way though the sumacs toward the south end of the little island. I stopped and drew breath as, through the grasses in a glade at the south end of the island, I saw a second adult crane and two little ones, about half as tall as the parents. Just standing there, wary but unperturbed. Incredible.

And it was somewhere, sometime about then that it struck me that this wasn't so much a bit of birding as it was a moment — a *moment* — in the evolution of the story of Craig Macdonald's map. I wasn't really sure why I'd made a point of coming here. At one time, I'd thought this lake might have been on the route that Larry Turner and Andrew Couchie took from Makõminising to Sawgidjeewayawgamawk, but by now I knew that they had gone south and west from Obabika and not north through here. But I *had* come here, to see the rare petroglyphs and, apparently, to meet this family of sandhill cranes. To meet these n'Daki Menan messengers who had something to say. But what?

After my trip, poking around a little bit through what the scientist in me had often discarded as New Age claptrap about animal totems and the like, I found a web page called "Anishinaabe Teachings of the Crane (Ajijaak)," which said this: "One of the Seven original clans (Dodem), the Crane (Ajijaak) represents wisdom and equal communication for all Anishinaabe. The Crane (Ajijaak) is also known as Baaswenazhii (Echo Maker) because they would sound their voices to gather our people for Ceremony, Council, and Celebrations. We are reminded by the Crane (Ajijaak) to think carefully and thoughtfully before we speak, and to do so in a clear and honest manner. As Anishinaabe, we believe that the Crane (Ajijaak) is the one who all of our Clans look to for guidance."[11]

And then it dawned on me: That's Craig. Craig Macdonald is the Baaswenazhii, the Echo Maker. It is he who listened to the Elders and created a physical living portrait — an echo — of knowledge they would surely take to the grave. The map represents wisdom and equal communication for all Anishinaabe. The map gathers people for Ceremony, Council, and Celebration. The map reminds us all to think carefully and thoughtfully before we speak and to do so in a clear and honest manner. *Meegwetch*, Ajijaak.

The lingering moment of this solo canoe journey through n'Daki Menan is sitting on a campsite on Temeeaygaming watching the sun set.

The haunting wail of a summer loon in the stillness drew my eyes to mature red and white pines of n'Daki Menan silhouetted on the skyline reserve. One classic white pine stood head and shoulders above the rest. And, for a moment, although Craig lived vicariously and intensively for nearly two years as I researched and wrote this book, I was glad the man himself was home in Dwight with Doris and not here. Because if he had been here with me, I wouldn't be sitting here. I would likely be traipsing along behind him in the gathering gloom to get to that massive pine. Just because. Maybe to see if one of the playful spirits of the place had somehow conspired for him to find his axe, tipped up against the base of the trunk. Some can't see the forest for the trees. Some can't see the trees for the forest. And some see one tree and walk toward it. Just because.

Acknowledgements

IN ACCEPTING THE INVITATION TO WRITE THE STORY OF CRAIG AND his map, I had two main goals in mind: The first was to try something a little different, in the wake of the findings of the Truth and Reconciliation Commission, and to collaborate with the people of Bear Island to tell the story of one of their heroes; and second, I hoped that the momentum of this book in telling Craig's story would generate interest to publish his other life's work, his book about winter travel technologies and techniques called *Traditional Sledding in North America*. I'm pleased to say that as of February 2025 that project is well underway with its new publisher, the Canadian Canoe Museum.

Working with the Teme-Augama Anishnabai, the Temagami First Nation, the Temagami Community Foundation, and Dundurn Press to create this book has been a privilege. I'm particularly proud (maybe not of the nearly three years it took) of the way we collaborated to protect the ownership of Traditional Indigenous Knowledge and, in doing so, found a way to rethink what a publishing contract might look like in a post-TRC world. I hope this process serves as a model for others who follow in this spirit of cross-cultural collaboration. And I thank everyone involved for their patience and good-heartedness in bringing this unique story to light.

Special thanks to Craig and Doris, of course, who welcomed me into their home and offered their individual perspectives on the lifetime of research that they jointly have undertaken over the last forty-plus years. Thank you as well to Craig's sister, Sandra, in New Zealand and to the Macdonald children — Colin, Janet, and Nancy — who shared with me their loving takes on their dad's many passions.

Appreciation to everyone in Temagami and beyond who also offered their insights into how the *Historical Map of Temagami* came to be, and why it is and will remain an appreciating gift for all time from the Elders who gave, through Craig, so willingly. To Richard and Victoria Grant, who championed this project from the very beginning, and who also opened their hearts and homes (in Newcastle and Temagami), you have inspired me at every turn with your foresight, kindness, and commitment.

And, finally, to my spouse and partner in all things, Gail Simmons: Let me be crystal clear in saying that none of this — the trips, the books, the hare-brained schemes, the backcountry shenanigans … the laundry — would have happened without your unconditional love and support at every turn.

JR
Cranberry Lake
February 2025

Appendix: Co-Creators of the Historical Map of Temagami

Peter Albany, Mowat's Landing, Ontario
Tom Alexander, Magnetawan, Ontario
Guy Alikut, Arviat, Nunavut
Louis Angalik, Arviat, Nunavut
Orrie Avery, Dorset, Ontario
Joseph Bananish, Longlac #58, Longlac, Ontario
Tonnice Batisse, Matachewan, Ontario
Frank Beaucage, Sturgeon Falls, Ontario
George Becker, Bear Island, Ontario
Walter Becker, Bear Island, Ontario
Mike Beedell, Ottawa, Ontario
Mike Bernard, Golden Lake, Ontario
Ralph Bice, Kearney, Ontario
Clarence Bogues, Dwight, Ontario
David and Anna Bosum, Oujé-Bougoumou, Quebec
Andy Buckman, Anima Nipissing Lake, Ontario
Robert Burns, South Portage, Ontario
Mike Buss, Dorset, Ontario
Roscoe Chapin, Unionville, Ontario
John Commanda, Garden Village, Sturgeon Falls, Ontario

Phillip Commanda, Golden Lake, Ontario
William Commanda, Kitigan Zibi, Quebec
George (Thor) Conway, Echo Bay, Ontario
Len Cote, Marten River, Ontario
Andrew Couchie, Temagami Island, Ontario
Ernest Couchie, Yelleck Point, North Bay, Ontario
John Couchie, Yelleck Point, North Bay, Ontario
Fred Courvoisier, Poverty Bay, Magnetewan, Ontario
William Cromarty, Chisasibi, Quebec
Bob Davis, Maple Lake, Haliburton, Ontario
Will (White Bear) Dedan, Fredericton, New Brunswick
Andrew Desmoulin, Marathon, Ontario
Luke and Gertie Diamond, Waskaganish, Quebec
Armand Duchesne, Pointe-Bleue, Lac-Saint-Jean, Quebec
Aubrey Dunne, Huntsville, Ontario
John and Christine Einish, Kawawachikamach, Schefferville, Quebec
Sampson Einish, Maliotenam, Quebec
Jim and Susan Espaniel, Levack, Ontario
Joe Evyagotailak, Kugluktuk, Nunavut
John Ojeeg Fisher, Garden Village, Sturgeon Falls, Ontario
Johnie Flaherty, Iqaluit (formerly of Grise Fiord), Nunavut
Lee Fletcher, Sault Ste. Marie, Ontario
Jean-Marie Fortin, Schefferville, Quebec
Robert Francis, Fort McPherson, Northwest Territories
Gus Friday, Temagami, Ontario
Sophie Friday, North Bay, Ontario
Don Galbraith, Fort Francis, Ontario
Oliver Glanfield, Fort Chipewyan, Alberta
Joe Goudie, Happy Valley, Labrador
Joe Guanish, Kawawachikamach, Quebec
Gordon Guppy, Temagami, Ontario
Adam and Agnes Haymond, Kipawa, Quebec
Karl Hoffman, Fort Smith, Northwest Territories
Flora Hunter, Kipawa, Quebec

Nancy Jones, Bear's Pass, Fort Frances, Ontario
Bill Katt, Bear Island, Ontario
Johnny Katt, Bear Island, Ontario
Sydney Keighley, Winnipeg, Manitoba
Pi Kennedy, Fort Smith, Northwest Territories
Denis LaRonde, Chimo Island, Lake Temagami, Ontario
Jim and Sarah Lavalley, Golden Lake, Ontario
Philias LePage, Elk Lake, Ontario (interviewed 1978 at age 108 and lived to 112)
Tom Linklater, Capreol, Ontario
John Mamianskum, Whapmagoostui (Great Whale River), Quebec
Alec Marshall, Pickerel River, Ontario
Elizabeth Mathias, Bear Island, Ontario
Wilfred McBride, North Temiskaming (Saging), Quebec
Ben McConnell, Deux Rivières, Ontario
Emmerson McConnell, North Cobalt, Ontario
Jack McConnell, Deux Rivières, Ontario
Jane (Wawataysi) McKee, Toronto, Ontario
Angus McKenzie, Kipawa, Quebec
Donald and Rosie McKenzie, Bear Island, Ontario
Dalton McLaren, South Lorraine, Ontario
Elizabeth (Kitty) McLaren, South Lorraine, Ontario
Fred McLeod, Garden Village, Sturgeon Falls, Ontario
Elliott Merrick, Asheville, North Carolina
Alex (Shimmy) and Tilley Misabi, Bear Island, Ontario
Angelique Misabi, Bear Island, Ontario
Frank Mitchell, L'Anse aux Meadows, Newfoundland
Ted Mongrain, Kipawa, Ontario
Harry Montague, North West River, Labrador
Alice Moore, North Bay, Ontario
Ed Moore, Cobalt, Ontario
Kieran Moore, Peterborough, Ontario
Sonny Moore, Bear Island, Ontario
Alphonse Peekongay Moses, Heron Bay, Ontario

Harry Muckwabi, French River, Ontario
Rick Nash, Dorset, Ontario
Guy Newell, Saint Anthony, Newfoundland
Charlie Neyelle, Délı̨nę, Northwest Territories
Dr. Jill Oakes, University of Manitoba, Winnipeg, Manitoba
Dorthe Olsen, Sisimiut and Kangerlussuaq Museums, Sisimiut, Greenland
Rita (Bubsee) O'Sullivan, Bear Island, Ontario
Leo (Mompey) Paul, Bear Island, Ontario
Michael Kush Kush Paul, Bear Island, Ontario
Joe Penasse, Toronto, Ontario
Bob Petrant, North Bay, Ontario
David Peunish, Constance Lake, Ontario
Jack Pierce, Elk Lake, Ontario
Mark Pimlott, Dwight, Ontario
Dan Pine, Garden River, Sault Ste. Marie, Ontario
Fred Pine, Garden River, Sault Ste. Marie, Ontario
George and Charlotte Pishabo, Bear Island, Ontario
Gary Potts, Bear Island, Ontario
Phillip Potts, Bear Island, Ontario
Pat Reynolds, Kipawa, Quebec
Ruby Robinson, Kawawachikamach, Quebec
Seymor Robinson, Ville-Marie, Quebec
Dr. Ed Rogers, Stoney Creek, Ontario
Charlie Ross, Wahnapitae, Ontario
Johnny Rowan, Tasso Lake, Ontario
George Saint Onge, Maliotenam, Quebec
Dan Sarazin Sr., Golden Lake, Ontario
Tom Saville Jr., North Bay, Ontario
Oscar Sawyer, Dorset, Ontario
Joe and Francine Shiner, Duck Lake, Chapleau, Ontario
Wellington Simcoe, Rama, Ontario
Gerard Simeon, Pointe-Bleue, Lac-Saint-Jean, Quebec
Frank Slade, Saint Anthony, Newfoundland
Doug Smarch, Teslin, Yukon Territory

Bill Smith, Temagami, Ontario
Peter Stanger, North Temiskaming (Saging), Quebec
Joe Stephens, Sturgeon Falls, Ontario
Alfred Stewart, Cache Bay, Ontario
Hugh Stewart, Wakefield, Quebec
Joe Stocking, Dorset, Ontario
Cote Taybishkwon, Letang, Quebec
Abel Taylor, Raleigh, Newfoundland
Madeline (Wachigo) Theriault, North Bay, Ontario
Ron Trolove, Stirling Falls, Ontario
George (Boychee) Turner, Shannon's Island, Lake Temagami, Ontario
Gordon Turner, Bear Island, Ontario
Bill Twain, Bear Island, Ontario
Dick Twain, Camp Wanapitei, Lake Temagami, Ontario
Arthur Twomey, Fort Lee, New Jersey
Henri Vaillancourt, Greenville, New Hampshire
Father Veyrat, St. Michael's Mission, Ross River, Yukon Territory
Stanley Wabase, Summer Beaver (Nibinamik), Ontario
Frank Wabi, North Temiskaming (Saging), Quebec
Alma Wallis, Rolphton, Ontario
Fred Wheatley, Parry Sound, Ontario
Mike Yellowhead, Collins, Ontario

Notes

Introduction: Binaakwe Geesis — Ma-kõmin-ising, Te-mee-ay-gaming

1 Cecelia Rose LaPointe, "Anishinaabe Star Knowledge Map & Stories" (blog post), September 23, 2014, archived August 25, 2017, at the Wayback Machine, web.archive.org/web/20170825230108/http://www.anishinaabekwe.com/blog/ishinaabekwe.com/2014/04/anishinaabe-star-knowledge-map-stories.html.

2 *The Silent Enemy*, directed by H.P. Carver, written by W. Douglas Burden and Richard Carver (August 2, 1930), youtube.com/watch?v=SK_S6GMbqfk.

3 Madeline Katt Theriault, *Moose to Moccasins: The Story of Ka Kita Wa Pa No Kwe* (Toronto: Natural Heritage/Natural History, 1992), 89.

4 This story comes from multiple sources, including archaeologist Thor Conway's "Temagami Secrets" Facebook page (facebook.com/profile.php?id=100063673022984).

5 The original radio broadcast from October 30, 1938, is captured in many places, including youtube.com/watch?v=Xs0K4ApWl4g&t=164s.

Part 1: Wabanong (East) — Beginnings
Mr. Macdonald Testifies in the Supreme Court of Ontario

1 Trial Proceedings, vol. 22, 4041–51.

2 Trial Proceedings, vol. 22, 4071–72.

3 Trial Proceedings, vol. 24, 4351–52.

Lifeline I — The Early Years

1 Sandra Macdonald, from the transcript of a conversation with the author from Macdonald's home in New Zealand, June 21, 2021.

2 Sandra Macdonald, transcript, June 21, 2021.
3 Accounts of conversations with Craig are taken, for the most part, from a series of recorded conversations in 2020–21 conducted at Craig's home in Dwight, Ontario, that were transcribed and archived as part of the research for this book.

Travelling with Craig I — Kandalore Days

1 One of the longest and most comprehensive things Craig has written is a monograph report on this trip titled "My Memory of the Lake Superior 1967 Centennial Canoe Voyage," published in the fall 2018 issue of *Nastawgan*, the Wilderness Canoe Association's quarterly journal.
2 Macdonald, "My Memory," 2.

Part 2: Shawanong (South) — Growth

Mr. Macdonald Continues Testimony in the Supreme Court of Ontario

1 Trial Proceedings, vol. 22, 4074–79.
2 Trial Proceedings, vol. 22, 4085–86.
3 Trial Proceedings, vol. 22, 4101–2.
4 Trial Proceedings, vol. 22, 4106–7.
5 Trial Proceedings, vol. 22, 4108–9.
6 Trial Proceedings, vol. 22, 4147.
7 Trial Proceedings, vol. 22, 4199–201.

Lifeline II — Taking Flight

1 Sandra Macdonald, transcript, June 21, 2021.
2 The contrast between Craig and his brother, Rod, couldn't be more stark. Both with keen and fertile minds, but where Craig was hands-on and rooted in the practical, Rod was the academic intellectual, as celebrated in a book published after his death in 2014 called *The Unbounded Level of the Mind: Rod Macdonald's Legal Imagination*, edited by Richard Janda, Rosalie Jukier, and Daniel Jutras (Montreal: McGill-Queen's Unversity Press, 2015).
3 Doris Macdonald, from the transcript of a conversation with the author by Zoom from the Macdonald home in Dwight, May 14, 2021.
4 Doris Macdonald, transcript, May 14, 2021.
5 George Moroz, from the transcript of a Zoom conversation with the author, June 16, 2021.
6 Craig Macdonald, from the transcript of a conversation with the author, May 12, 2020.
7 Doris Macdonald, transcript, May 14, 2021.

Travelling with Craig II — James Bay Bound

1 Craig Macdonald, transcript, May 13, 2020.
2 A reminder of these details comes from Craig's memory, supplemented by an article called "Postmen of the Far North," from *Canada: An Illustrated Weekly Journal for All Interested in the Dominion* 9 (January 25, 1908).
3 After the trip, Craig directed me to his source for some of this story, which was an article called "Pritchard and Lagimodière" by Clifford Wilson in *The Beaver*, June 1948, 18–21.
4 For another account of this trip, see my article "Tumpline Trek on James Bay Shores," *Canadian Geographic* 102, no. 6 (1982): 54–59.

Part 3: Ningaabii'anong (West) — Finding Meaning

Mr. Macdonald Is Cross-Examined by T. McCabe, Counsel for the Plaintiff

1 Trial Proceedings, vol. 24, 4448–64.

Lifeline III — The Map Comes to Be

1 Nancy Macdonald, from the transcript of a conversation between the author and the Macdonald siblings, April 21, 2021.
2 Doris Macdonald, from the transcript of a Zoom conversation with the author, May 19, 2021.
3 Craig Macdonald, "The Nastawgan: Traditional Routes of Travel in the Temagami District," in *Nastawgan: The Canadian North by Canoe and Snowshoe*, ed. Bruce W. Hodgins and Margaret Hobbs (Toronto: Betelgeuse Books, 1985), 183.
4 Craig Macdonald, "Nastawgan," 188.
5 Erhard Kraus, "Kanu Links," August 1999, interlog.com/~erhard/nastagan.htm (site discontinued).
6 From a redacted excerpt of the minutes from the relevant meetings of the Ontario Geographic Names Board, provided by the Ontario Ministry of Natural Resources in 2022.
7 There were at the time maps of Indigenous place names and routes produced by various land claim–related initiatives, including preparatory research for the Labrador Inuit Land Claims Agreement, the Inuit Land Use and Occupancy Project, and the Dene Mapping Project (which became key documents in the division of the Northwest Territories and the creation of Nunavut). But these did not have the level of detail that Craig had achieved for the homeland of the Teme-Augama Anishnabai.

8 A detailed consideration of Speck's work in Temagami is in a wonderful collection of digitized media and texts housed at temagami.nativeweb.org.
9 Not her real name.

Travelling with Craig III — In the Land of the Montagnais

1 Colin Macdonald, from the transcript of a Zoom conversation between the author and the Macdonald siblings, April 22, 2021.
2 This is a direct reference to problems with Jim Greenacre bolting ahead on the latter days of the James Bay trek.
3 Tom Linklater's memoir, *The Last Forest Ranger: Algonquin Park Memories*, which has many other heartwarming and often funny stories, was published by the Friends of Algonquin Park in 2021.
4 Not his real name, which was used in the original letter.
5 Accompanying Craig on this trip were Dr. Bill King, who had been with him on the James Bay trek and in Jacques Cartier Provincial Park, and Tony Bird. The fourth member of the expedition was legendary backcountry traveller Herb Pohl, now departed, whose book *The Lure of Faraway Places: Reflections on Wilderness and Solitude* (Toronto: Natural Heritage Books, 2007) was an important addition to the writings about wilderness travel.
6 This trek is described by Herb Pohl in "Clearwater Winter, 1987" in the winter 2014 issue of *Nastawgan*.
7 From a recorded Zoom conversation with Doris Macdonald on Friday, May 14, 2021.
8 Janet became a chemistry professor at Vanderbilt University in Nashville, Tennessee. These recollections come from a long letter she wrote to me about her dad after we spoke on the phone.
9 From a letter from Janet Macdonald.

Part 4: Keewatinong (North) — Legacy

Lifeline IV — Moving On

1 Daughter Janet reflected on the origin of her dad's legendary physicality in the woods: "My grandmother (Dad's mom) was an INCREDIBLE athlete. The cottage is littered with her trophies and ribbons from track and field events. She was the pitcher for the Ontario champion baseball team. Dad always said no kid on their street could hit her pitches." I mentioned this because Craig-o-matic got that athleticism from somewhere. It was his mom.
2 Which is the way things remain to this day, although Craig does like to walk into the kitchen when Doris has her computer nearby and say, "Alexa, what

is the value of pi?" This, of course, prompts Alexa to go on and on reciting her answer to dozens of decimal places, which Craig thinks is hilarious.

3 Doris Macdonald, transcript, May 14, 2021.

4 Colin Macdonald, transcript, April 22, 2021.

5 Janet Macdonald, from the transcript of a Zoom conversation between the author and the Macdonald siblings, April 22, 2021.

6 Nancy Macdonald, from the transcript of a Zoom conversation between the author and the Macdonald siblings, April 22, 2021.

7 Nancy Macdonald, transcript, April 22, 2021.

8 For a detailed account of this situation, see Jocelyn Thorpe, *Temagami's Tangled Wild: Race, Gender, and the Making of Canadian Nature* (Vancouver: UBC Press, 2012), 80–81.

9 Andy Buckman, from the transcript of a telephone conversation with the author, March 10, 2022.

10 From an interview with the author in Hugh Stewart's cabin at Camp Temagami on Temeeaygaming, July 14, 2021.

11 Keith Helmuth, *Tappan Adney: And the Heritage of the St. John River Valley* (Woodstock, NB: Chapel Street Editions, 2017).

12 E. Tappan Adney Papers MG H22 Case 3, File 5, page 6, University of New Brunswick Archives, as cited in James W. Wheaton, "Tappan Adney's Maliseet Studies: More Than Canoes" (presentation, 34th Algonquin Conference on Language and Linguistics held at Queen's University, Kingston, October 24–27, 2002).

13 These models are gloriously celebrated in John Jennings's coffee-table book *Bark Canoes: The Art and Obsession of Tappan Adney* (Toronto: Firefly Books, 2012).

14 It was first published as Bulletin 230 of the Smithsonian Institution and subsequently republished several times by the Smithsonian and other publishers.

15 William Arthur Allen, "2006 Algonquin Park Cultural Heritage Research Report Under Grandfather Teaching Theme: Wisdom and Ontario Archaeology Licence A070" (unpublished manuscript, September 15, 2006), 108.

16 Thorpe, *Temagami's Tangled Wild*, 94.

17 Quotes from Victoria and Richard Grant are taken from an interview with the author on Deacon's Island, Temeeaygaming, July 15, 2021.

Travelling with Craig IV — Rock Star Return to Bear Island

1 Joseph Burr Tyrrell, ed., *David Thompson's Narrative of His Explorations in Western America, 1784–1812* (Toronto: Champlain Society, 1916), i.

2 Ronald Rees, *New Brunswick Was His Country: The Life of William Francis Ganong* (Halifax: Nimbus Publishing, 2016).
3 Hugh Stewart, *Canoe Trails and Shop Tales: Making Crooked Nerves Straight* (Ottawa: McGahern Stewart Publishing, 2018).
4 Happily, in a continuing effort to regain cultural control of their lives in the lands of n'Daki Menan since the trial, the leadership of Temagami First Nation and Teme-Augama Anishnabai have commissioned many studies and reports that the Elders may not have seen, including a document titled *Dams on n'Daki Menan*, prepared by Jamie Friday, Temagami First Nation Lands and Resources, dated July 7, 2015, and filed on the Temagami First Nation website: temagamifirstnation.ca/wp-content/uploads/2017/10/Dams-on-nDaki-Menan-Draft.pdf.
5 Brandi Lee MacDonald, "Methodological Developments for the Geochemical Analysis of Ochre from Archaeological Contexts: Case Studies from British Columbia and Ontario, Canada" (Ph.D. diss., McMaster University), macsphere.mcmaster.ca/bitstream/11375/18635/2/MacDonald_Brandi_L_2015Dec_PhD.pdf.

Afterword

1 Mary Laronde, from the transcript of a conversation with the author in Laronde's home in North Bay, November 4, 2021.
2 David James Twain, November 14, 1941, to July 29, 2013. Age seventy-one years at his death. Son of Bill and Isobel Twain.
3 June Twain, from a conversation with the author at Bear Island, November 3, 2021.
4 Hap Wilson, from the transcript of a conversation with the author at Cabin Falls Ecolodge, July 6, 2021.
5 Hap Wilson, email to the author, November 16, 2021.
6 Wilson, transcript, July 6, 2021.
7 Throughout the research for this book, the work of author and historian Brian Back has been uniquely valuable. His article about Alex Mathias is part of an extremely informative website, Ottertooth.com.
8 From the Northwaters & Langskib brochure available with other tourist information at the Temagami Train Station store.
9 Andy Buckman, from the transcript of a recorded telephone conversation with the author on Thursday, March 10, 2022.
10 Alex Mathias, from the transcript of a conversation with the author on Lake Obabika, July 11, 2021.

11 Sault Tribe of Chippewa Indians, "Anishinaabe Teachings of the Crane (Ajijaak)," n.d., saulttribe.com/images/downloads/government/tribal%20code/Anishinaabe_Crane_Teachings.pdf.

Image Credits

xiii	Craig Macdonald
2	(top) Peter Couchie; (bottom) Eva Couchie
18–19	Craig Macdonald
34	Macdonald family
37	Macdonald family
46	Macdonald family
99	Macdonald family
106	Mary Katt/Courtesy of Mae Katt
117	Craig Macdonald/Photo by James Raffan
122	James Raffan
137	Teme-Augama Anishnabai
142	James Raffan
143	James Raffan
145	(top and bottom) James Raffan
158	Hugh McKenzie
161	Donald B. Smith
180	Craig Macdonald
184	Craig Macdonald
215	Macdonald family
234	Macdonald family
241	(top) Teme-Augama Anishnabai/Thor Conway; (bottom) Alex Paul
253	Macdonald family
285	Craig Macdonald
289	Alex Mathias

About the Author

Gail C. Simmons

Dr. James Raffan is a prolific writer, speaker, and geographer who has a long association with the Canadian Canoe Museum in Peterborough, Ontario. Over the years he has produced a number of bestselling books, including his critically acclaimed most recent title, *Ice Walker: A Polar Bear's Journey Through the Fragile Arctic*, which has been translated and republished in France and Spain. He has also written for media outlets including *Canadian Geographic*, *National Geographic*, and *The Globe and Mail*, as well as for CBC Radio and the Discovery Channel. James is a Fellow International of the Explorers Club, past chair of the Arctic Institute of North America, and was named in 2017 as one of Canada's most influential explorers of all time by *Canadian Geographic*. Recognitions include the Queen's Golden and Diamond Jubilee Medals, the Royal Canadian Geographical Society's Camsell Medal, and Canada's Meritorious Service Medal, all for service to Canada and its communities.

The Historical Map of Temagami

by Craig K. Macdonald

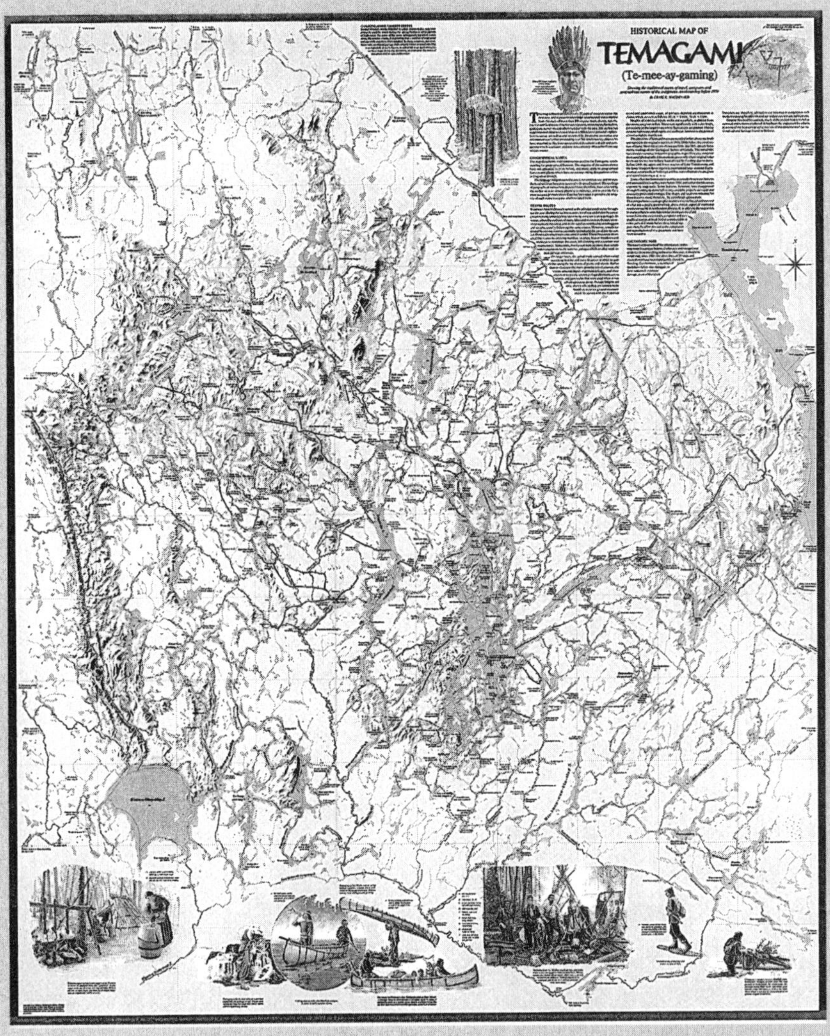

Full colour map • 38 x 58" • $50.00 plus shipping

Available from: Craig Macdonald, macdonal@vianet.ca, (705) 635-3416